Series in Real Analysis – Vol. 14

Nonabsolute Integration on Measure Spaces

SERIES IN REAL ANALYSIS

ISSN: 1793-1134

Published

Series in Real Analysis – Vol. 14

Nonabsolute Integration on Measure Spaces

Ng Wee Leng

Nanyang Technological University, Singapore

World Scientific

NEW JERSEY · LONDON · SINGAPORE · BEIJING · SHANGHAI · HONG KONG · TAIPEI · CHENNAI · TOKYO

Published by

World Scientific Publishing Co. Pte. Ltd.
5 Toh Tuck Link, Singapore 596224
USA office: 27 Warren Street, Suite 401-402, Hackensack, NJ 07601
UK office: 57 Shelton Street, Covent Garden, London WC2H 9HE

Library of Congress Cataloging-in-Publication Data
Names: Ng, Wee Leng.
Title: Nonabsolute integration on measure spaces / Wee Leng Ng (NTU, Singapore).
Description: New Jersey : World Scientific, 2017. | Series: Series in real analysis ; vol. 14
Identifiers: LCCN 2017005549 | ISBN 9789813221963
Subjects: LCSH: Numerical integration. | Integrals. | Henstock-Kurzweil integral. |
    Integrals, Generalized. | Algebraic spaces.
Classification: LCC QA308 .N4 2017 | DDC 515/.43--dc23
LC record available at https://lccn.loc.gov/2017005549

British Library Cataloguing-in-Publication Data
A catalogue record for this book is available from the British Library.

Printed in Singapore

To my beloved parents

Foreword

The major contribution of this book is proving the Radon–Nikodým theorem for the Henstock–Kurzweil integral on measure spaces with metric topologies which include the n-dimensional Euclidean space as a special case.

As it is known, the Radon–Nikodým theorem provides an alternative presentation of derivatives for the Lebesgue theory. It is now valid for the Henstock–Kurzweil integral which is a nonabsolute integral. By a nonabsolute integral, we mean that an integrable function needs not be absolutely integrable.

Other standard results of the Henstock–Kurzweil integral are also given, including the equi-integrability of Jaroslav Kurzweil and the controlled convergence theorem proved originally in 1951 by A. G. Djvarsheishvili [8] and in the language of Henstock in 1985 by Lee Peng Yee and Chew Tuan Seng [26].

The original definition of the (special) Denjoy integral was given in terms of Cauchy–Harnack extension and transfinite induction. It is called Harnack extension in this book. The Harnack extension is known to be a real-line dependent property. Due to this fact, the book entitled Theory of the Integral by Saks in 1937 [41] has served as a major reference for nonabsolute integration for many decades. A metric space analogue, and hence an n-dimensional Euclidean space version, of the Harnack extension is now available in this book. The results presented and proof techniques employed also contain ideas that might be worthy of, and could motivate, further research on nonabsolute integration.

Lee Peng Yee
July 2017

Preface

Theory of integration is the foundation of real analysis. There are various definitions of the integral, such as Riemann, Lebesgue, Denjoy, and Perron, just to name a few. The Riemann integral is applied essentially to continuous functions and the Lebesgue integral to absolutely integrable functions. The other integrals are not popular in calculus or real analysis courses as their definitions are either rather complex or demand different prerequisites from those for the Riemann integral and the Lebesgue integral. Many different notions of the integral were introduced in the last century, for real-valued functions, in order to generalise the Riemann integral. A discussion of the various definitions of the integral for real-valued functions can be found in [19].

In the fifties of the twentieth century, Ralph Henstock and Jaroslav Kurzweil discovered and developed independently an integral, now commonly known as the Henstock–Kurzweil integral, which includes both the Riemann integral and the Lebesgue integral on the real line, and is elegantly similar in its form to the Riemann integral. Furthermore, the Henstock–Kurzweil integral is nonabsolute, that is, there are functions which are Henstock–Kurzweil integrable but whose absolute values are not, and it integrates all derivatives of differentiable functions. The simplicity of the definition of the Henstock–Kurzweil integral, coupled with the perceived power and utility of the integral, has led some mathematics educators to advocate that this integral be taught in introductory analysis courses instead of the Lebesgue integral.

Given the potential of the Henstock–Kurzweil integral to advance our knowledge on the theory of integration, in this book we shall present an original theory of a Henstock–Kurzweil type integral, which we call the H-integral, on measure spaces endowed with locally compact metrizable topologies that are compatible with the measure. We will show that the H-integral is nonabsolute, and indeed a generalisation of the Henstock–Kurzweil integral, deal with the essential aspects of integration theory and prove results which could have further applications.

The key tools used in defining the Henstock–Kurzweil integral on the Euclidean space, in particular the real line, include the notions of intervals and gauges. A gauge, usually denoted by δ, is a positive function defined on the domain of integration which plays an important role in partitioning the latter into a finite collection of interval-point pairs known as a δ-fine division. As general measure spaces or topological spaces do not have any properly defined intervals, we will first introduce the notion of generalised intervals, in measure spaces with a suitable topological structure, which is a generalisation of intervals on the real line, and define G-fine divisions, where G, also called a gauge in this book, is to the H-integral what δ is to the Henstock–Kurzweil integral. More specifically, we will describe explicitly how certain objects in measure spaces endowed with locally compact metrizable topologies are chosen to be generalised intervals, and prove that given any gauge G, a G-fine division exists.

This book contains both results that are fundamental to any Henstock–Kurzweil type integral such as Henstock's lemma and convergence theorems as well as more advanced results such as equiintegrability theorem, the Radon–Nikodým theorem, and Harnack extension. Complete proofs of all the key results are provided in this book so as to exhibit the various techniques typically used in proving results in theories of integration. A synopsis which follows this preface provides an outline of what we will discuss in each of the five chapters of this book. For the reader's convenience in recalling the key definitions and results presented, in addition to an index, a glossary is also provided at the end of this book.

I wish to take this opportunity to express my gratitude to Professor Lee Peng Yee for being my source of inspiration all these years. It is my honour that he so graciously wrote the foreword for this book.

Ng Wee Leng
August 2017

Synopsis

In Chapter 1, we define the H-integral on an elementary set, which is a finite union of mutually disjoint generalised intervals, and establish properties that are fundamental to an integral. We describe explicitly in Section 1.1 how certain objects in measure spaces endowed with locally compact metrizable topologies are chosen to be generalised intervals and elucidate the concept of generalised intervals using several concrete examples including the Euclidean space. The existence of a G-fine division given any gauge G, when the topology is metrizable, is proved before the H-integral is defined in Section 1.2. In addition, we prove that the H-integral includes the Henstock–Kurzweil integral on the real line. To build the fundamental theory for further development of the H-integral, basic but essential properties of the H-integral, in particular, the Henstock's lemma and the monotone convergence theorem, are proved in Section 1.3.

Chapter 2 aims to relate the H-integral to some known integrals. In Section 2.1, we define the M-integral, which is a McShane-type integral, and prove that a function is M-integrable on an elementary set E if and only if it is absolutely H-integrable on E. This result strongly supports the fact that the H-integral is a generalisation of the Henstock–Kurzweil integral to measure spaces. The domains of H-integrability and M-integrability are also extended from elementary sets to measurable sets. Subsequently, in Section 2.2, we establish the equivalence between the M-integral and the Lebesgue integral which supports the validity of the M-integral. We also show that a function which is H-integrable on an elementary set is Lebesgue integrable on a portion of the elementary set. In Section 2.3 we establish the fact that the H-integral includes the Davies integral as well as the Davies–McShane integral defined by Henstock in [15]. This is done by establishing the equivalence between the Lebesgue integral and the Davies integral, as well as that between the Lebesgue integral and the Davies–McShane integral. The chapter concludes with the result that for measurable functions, the absolute H-integral, the M-integral, the Lebesgue integral, the Davies integral, and the Davies–McShane integral are all equivalent.

Further results of the H-integral are given in Chapter 3. We begin by proving in Section 3.1 that H-integrable functions are measurable, and proceed to give a necessary and sufficient condition for a function to be H-integrable. We also prove that the H-integral is genuinely a nonabsolute one by defining explicitly a function which is H-integrable but not absolutely H-integrable. Three concepts very relevant to H-integrability, namely the generalised absolute continuity, H-equiintegrability, and the strong Lusin condition, are introduced in Section 3.2 and key results involving these concepts are proved. Section 3.3 is devoted to proving the key convergence theorems for the H-integral. We begin with the proofs of the equiintegrability theorem and the basic convergence theorem, and illustrate how the generalised mean convergence theorem can be proved with the aid of the two aforementioned convergence theorems. The controlled convergence theorem is subsequently proved in a few lemmas and by applying the basic convergence theorem.

Chapter 4 is the highlight of this book as it presents the most important results we have proved for the H-integral on measure spaces endowed with a locally compact metrizable topology. Parallel results for the Euclidean space have been proved by the author and Lee Peng Yee in [28] and those for metric spaces by the author in [36]. It is well-known that, on the real line, there is a correspondence between the Henstock–Kurzweil integrable functions and the so-called ACG^* functions [21]. This correspondence is essentially the Radon–Nikodým theorem. However, for a long time, this result could not be extended to higher dimensional spaces because in the latter cases multiple integrals are involved. In this chapter we use an approach to generalise the Radon–Nikodým theorem that is to a certain extent inspired by the corresponding result for the Lebesgue integral. The main theorem, namely the Radon–Nikodým theorem for the H-integral, is proved in Section 4.1 with which we proceed to provide a descriptive definition of the H-integral in Section 4.2. By imposing a different condition, a second version of the main theorem, and subsequently a second descriptive definition of the H-integral are also given. The purpose of Section 4.3 is to present results which we prove in Section 4.1 for the HK-integral, which is a natural extension of the Henstock–Kurzweil integral to higher dimensional Euclidean spaces. We also show how some well-known results on the real line, for example, the fundamental theorem of calculus for the Henstock–Kurzweil integral, can be deduced from the descriptive definition of the HK-integral obtained.

In Chapter 5, we prove that the Harnack extension for the Henstock–Kurzweil integral on the real line is available for the H-integral on metric spaces which are locally compact. We also recover the proof by means of category argument, which is a standard approach for classical integration theory on the real line, and prove the Harnack convergence theorem which is a kind of dominated convergence theorem. In Section 5.1, we present the terminology for defining the H-integral using the language of metric spaces. Harnack extension for the H-integral will be proved in Section 5.2 after two important concepts, namely the δ-fine cover and the nonabsolute subset, that are pivotal to the notion of Harnack extension are presented. The Cauchy extension for the H-integral is obtained as a corollary of the Harnack extension. In Section 5.3, the notion of uniform Harnack condition is introduced and the Harnack converegnce theorem is then proved by means of the category argument and by applying the Harnack extension. This is done by first proving the dominated convergence theorem and the mean convergence theorem for the H-integral. In Section 5.4, we show that an improved version of the controlled convergence theorem for the H-integral can be proved by applying the Harnack convergence theorem for the H-integral. We end the chapter with a result which relates the uniform Harnack condition to the H-equiintegrability.

Contents

Chapter 1

A Nonabsolute Integral on Measure Spaces

Ralph Henstock developed a general theory of the integral [12; 14] which includes as special cases the Denjoy–Perron integral [23], the approximate Perron integral [4], the Weiner integral [45] and the Feynman integral [34], as well as the general Denjoy integral [42]. Here, we shall consider a special case of his theory. More precisely, we shall define a Henstock-type integral on measure spaces endowed with locally compact Hausdorff topologies, and show that in cases where the topologies are metrizable, the integral defined is a nonabsolute one.

To our knowledge, prior to the author's joint work with Lee [37], there were at least two attempts to define an integral in a similar setting. In [1], Ahmed and Pfeffer defined an integral in a locally compact Hausdorff space, and in [15], Henstock constructed a division space from an arbitrary non-atomic measure space with a locally compact Hausdorff topology that is compatible with the measure, and defined the Davies–McShane integral. The setting in [1] is different from that of our work as we aim to define an integral on measure spaces, albeit those with certain topological structures. The work in [15] is relatively closer to what we shall present in this chapter. However, the integral defined in [1] and that in [15] are both *absolute* in the sense that if f is integrable, then so is $|f|$, where $|f|$ denotes the absolute value of f.

The main theme of this chapter is thus to show how a non-absolute integral, which we shall call the H-integral, can be defined on measure spaces endowed with locally compact Hausdorff topologies. An account of how the H-integral is related to the Davies–McShane integral will be given in Chapter 2, and an example to show that the H-integral is nonabsolute will be given in Chapter 3.

1.1 Preliminaries

Throughout this book, we shall consider a *measure space* (X, Ω, ι), where $(X, \mathcal{T})$ is a topological space, Ω is a σ-algebra on X, ι is a measure on Ω, and $\mathcal{T}$ is a locally compact Hausdorff topology that is compatible with the measure ι in a sense which we will make precise later.

To make this book sufficiently self-contained, we review briefly here concepts related to set-theoretic topology which are needed in this book. Relevant concepts in measure theory will be reviewed as and when they are required.

First, members of the topology $\mathcal{T}$ are called *open sets* as usual. A set Y is *closed* if the *complement* of Y, denoted by $X \setminus Y$, is open. We also use $Y_1 \setminus Y_2$ to denote the *relative complement* or *difference* of two subsets Y_1 and Y_2 of X, that is, $Y_1 \setminus Y_2$ is the set of all points in Y_1 that are not in Y_2. Given $x \in X$, if U is an open set such that $x \in U$, then U is said to be an *open neighbourhood* of x. For every $Y \subseteq X$, we shall always denote its *closure* by $\overline{Y}$ and we say that Y is *relatively compact* if $\overline{Y}$ is compact. A set Y is *compact* if every open cover of Y has a finite subcover. An *open cover* of Y is a collection of open sets whose union contains Y. The *interior* and *boundary* of Y shall be denoted by Y^o and ∂Y respectively and we define $\partial Y = \overline{Y} \setminus Y^o$.

By *locally compact* we mean that every element in X has a relatively compact open neighbourhood, and by *Hausdorff* we mean any two distinct points in X can be separated by two disjoint open neighbourhoods. A set Y is *connected* if it is not the union of non-empty disjoint sets of the form $Y \cap U$ and $Y \cap V$, where U and V are open sets.

We remark that since in this book we assume that X is locally compact and Hausdorff, we have the property that for every $x \in X$ and for any open neighbourhood U of x, there exists a relatively compact open neighbourhood V of x such that $\overline{V} \subseteq U$. It follows that X is a *regular space* in the sense that if Y is a closed subset of X and $x \notin Y$, then x can be separated from Y by two disjoint open sets. Furthermore, the topology $\mathcal{T}$ of X has a *basis* $\mathcal{T}_1$ consisting of relatively compact open sets, that is, $\mathcal{T}_1 \subseteq \mathcal{T}$ is such that each open set is the union of members of $\mathcal{T}_1$ each of which is relatively compact. Also note that compact subsets of X are closed while closed subsets of a compact set in X are compact.

We highlight that the case where $\mathcal{T}$ is a locally compact metric topology is included in our setting as a special case. It is easy to see that metric spaces are Hausdorff. However, not all metric spaces are locally compact. For instance, the space of all rational numbers endowed with the topology from the space of all real numbers is not locally compact since all its compact sets have empty interiors and are therefore not neighbourhoods. The lower limit topology and the upper limit topology on the set of all real numbers are not locally compact either. On the other hand, metric topologies which are locally compact include the Euclidean space (and in particular the real line), topological manifolds, the Cantor set, the Hilbert cube, and all discrete spaces.

For a more detailed account of the aforementioned topological concepts and properties, we refer the reader to [9].

The remainder of this chapter is organised as follows. The necessary preliminaries for defining an integral will be given in the rest of this section. In particular, generalised intervals in measure spaces will be defined. In Section 1.2, the existence of a G-fine division will be proved before we proceed to define the H-integral. Standard properties of the H-integral that will build the fundamental theory for further development will be derived in Section 1.3.

One of the key tools in defining the Henstock–Kurzweil integral in the Euclidean space, and in particular on the real line, is the notion of intervals. To define a Henstock-type integral on a general measure space or topological space, in which there are no properly defined intervals, we need to identify suitable objects in the space as generalised intervals, and give a meaning to division. In order to obtain a nonabsolute integral, we have to be stringent in our choice of generalised intervals. In this section, we shall define the generalised interval and relate the definition to the real line, as well as some metric spaces, prior to presenting the standard definitions and terminology in defining a Henstock-type integral.

We reiterate that (X, Ω, ι) is a measure space, where Ω is a σ-algebra of subsets of X and ι is a measure on Ω, endowed with a locally compact Hausdorff topology $\mathcal{T} \subseteq \Omega$.

A *σ-algebra* on X is a collection of subsets of X that includes X, and is closed under complements and countable unions. It follows that a σ-algebra on X includes the *empty set*, denoted by $\emptyset$ throughout this book, and is closed under countable intersections. Each $W \in \Omega$ is called a *measurable set*. We emphasise that all open sets are measurable in our setting, as is the case in [15]. The *measure* $\iota : \Omega \longrightarrow [0, \infty]$ is a function satisfying $\iota(\emptyset) = 0$ which is *countably additive*, that is,

$$\iota\left(\bigcup_{i=1}^{\infty} W_i\right) = \sum_{i=1}^{\infty} \iota(W_i)$$

for any $W_i \in \Omega$, $i = 1, 2, \ldots$, which are pairwise disjoint sets. The following condition, which we shall hereafter refer to as Condition ($*$), will be assumed throughout this book.

For every measurable set W and every $\varepsilon > 0$, there exist an open set U and a closed set Y such that $Y \subseteq W \subseteq U$ and

$$\iota(U \setminus Y) < \varepsilon. \tag{$*$}$$

Obviously if the above inequality holds, then we also have $\iota(W \setminus Y) < \varepsilon$ and $\iota(U \setminus W) < \varepsilon$.

We shall next define the generalised interval. In [15], Henstock considered all measurable sets in his construction of a division, and as a result an absolute integral was obtained. This shows that in order to define a nonabsolute integral, we must recruit fewer objects as generalised intervals.

To motivate our argument, we observe that on the real line, bounded intervals are used to define divisions, and a bounded interval can be seen as the difference of two bounded intervals such that one does not contain the other. Generalising this concept to the more abstract setting of a measure space which has a topological structure, a generalised interval could be the difference of two connected sets such that neither is a subset of the other, so that their difference remains a connected set.

However, as we shall see in the proof of the existence of a division in Section 1.2, in order to construct a division with mutually disjoint generalised intervals, defining the set of generalised intervals as the collection of sets each of which is the difference of two connected sets such that neither is a subset of the other is not sufficient. To ensure the existence of a division, as we shall see, generalised intervals should also include finite intersections of sets which we just described.

Let us now define the generalised interval formally. Let $\mathcal{T}_1$ be a basis for $\mathcal{T}$ consisting of relatively compact open sets. As mentioned earlier, such a basis always exists as we assume that $\mathcal{T}$ is locally compact and Hausdorff. If the topology $\mathcal{T}$ of X is induced by a metric d on X, then $\mathcal{T}_1$ is the set of all d-open balls. A set of the form $\{y \in X : d(x, y) < r\}$ where $x \in X$ and $r > 0$, denoted by $B(x, r)$, is called a *d-open ball*, or simply *open ball*, with centre x and radius r. We shall also call its closure a *d-closed ball*, or simply *closed ball*. Throughout this book we shall assume that for all $U \in \mathcal{T}_1$, we have $\iota(U) > 0$, if $U \neq \emptyset$, and $\iota(U) = \iota(\overline{U})$. Consequently, we have $\iota(\partial U) = 0$ for all $U \in \mathcal{T}_1$.

Consider the following sets.
$$\mathcal{I}_0 = \left\{ \overline{U_1} \setminus \overline{U_2} : U_1, U_2 \in \mathcal{T}_1 \text{ where } U_1 \nsubseteq U_2 \text{ and } U_2 \nsubseteq U_1 \right\},$$
$$\mathcal{I}_1 = \left\{ \bigcap_{i \in \Lambda} V_i \neq \emptyset : V_i \in \mathcal{I}_0 \text{ and } \Lambda \text{ is a finite index set} \right\}.$$
Note that by definition, $\mathcal{I}_0$ includes all sets of the form $\overline{U}$, where $U \in \mathcal{T}_1$, and that $\mathcal{I}_0 \subseteq \mathcal{I}_1$. It is easy to see that $\mathcal{I}_1$ is closed under finite intersections, if the intersection is non-empty, and that since Ω is a σ-algebra, members of $\mathcal{I}_1$ are measurable.

We shall call each $I \in \mathcal{I}_1$ a *generalised interval*, or simply *interval* where there is no ambiguity. Note that generalised intervals are relatively compact, though not necessarily closed or compact. Also note that generalised intervals are connected. Since we assume that $\iota(U) = \iota(\overline{U})$ for each $U \in \mathcal{T}_1$, it follows that for each generalised interval I, the property that $\iota(I) = \iota(\overline{I})$ holds.

To elucidate the definition of a generalised interval, let us relate it to the real line.

Example 1.1   Let X be the real line $\mathbb{R}$ and $\mathcal{T}$ be the topology induced by the usual metric d, that is, $d(x,y) = |x - y|$, the absolute value of $x - y$, for $x, y \in \mathbb{R}$. Then $\mathcal{T}_1$ is the set of all intervals of the form (a, b). It is easy to see that $\mathcal{I}_0$ is the set of all intervals of the form $(a, b]$, $[a, b)$ or $[a, b]$ while $\mathcal{I}_1$ is the set of all intervals of the form (a, b), $(a, b]$, $[a, b)$ or $[a, b]$. Here we use the standard notations for intervals on the real line. For example, $(a, b) = \{x \in \mathbb{R} : a < x < b\}$ and $(a, b] = \{x \in \mathbb{R} : a < x \leq b\}$. In other words, generalised intervals in this case are the usual bounded intervals. Note that taking the difference of two bounded intervals such that one does not contain the other ensures that we obtain a connected interval rather than two disjoint bounded intervals.

In the next two examples we shall describe what generalised intervals look like when $\mathcal{T}$ is a locally compact metric topology. We first consider a general metric space, and then the two-dimensional Euclidean space.

Example 1.2   Suppose that the topology $\mathcal{T}$ of X is induced by a metric d on X and $\mathcal{T}$ is locally compact. As explained previously, this is a special case of our setting. Let $\mathcal{T}_1$ be the set of all d-open balls. The difference of two d-closed balls such that one does not contain the other is a typical member of $\mathcal{I}_0$. Thus, members of $\mathcal{I}_0$ are either d-closed balls or scalloped balls. Consequently, a generalised interval is a finite intersection of a combination of closed balls and scalloped balls such that the intersection is non-empty. We reiterate that taking the difference of two d-closed balls such that one does not contain the other ensures that we obtain a connected scalloped ball with no 'holes'.

We shall next use a more concrete example, namely the two-dimensional Euclidean space, to illustrate how the different choices of the metric can give rise to different generalised intervals.

Example 1.3 Let X be the two-dimensional Euclidean space $\mathbb{R}^2$. The metrics d_1 and d_2 on X are given by

$$d_1(x, y) = \max\{|x_1 - y_1|, |x_2 - y_2|\},$$

and

$$d_2(x, y) = [(x_1 - y_1)^2 + (x_2 - y_2)^2]^{\frac{1}{2}},$$

for each $x = (x_1, x_2)$ and $y = (y_1, y_2)$ in X. It is well known that d_1-open balls are squares without the boundaries, and d_2-open balls are open circular discs. With reference to Example 1.2, we can see that when the metric d_1 is used, a generalised interval looks like a polygon with edges each being either vertical or horizontal, and each edge is not necessarily included. When the metric d_2 is used instead, a generalised interval is a simply connected domain in the plane with edges being circular arcs, and each circular arc may or may not be included.

We next present the necessary and standard terminology in defining a Henstock-type integral.

Let E be a finite union of (possibly just one) mutually disjoint intervals, and call it an *elementary set*. Note that intervals are themselves elementary sets. Furthermore, since $\iota(I) = \iota(\overline{I})$ for each generalised interval I, we also have the property that $\iota(E) = \iota(\overline{E})$ for each elementary set E. An elementary set E is said to have a *finite measure* if $\iota(E) < +\infty$.

Throughout this book, we shall let an elementary set E with a finite measure be fixed, and define integrability on E.

If a subset E_0 of E is an elementary set, then E_0 is said to be an *elementary subset* of E. If both E_0 and $E \setminus E_0$ are elementary sets, then we call E_0 a *fundamental subset* of E. If $I \subseteq E$ and I is an interval, then as usual we call I a *subinterval* of E. A subinterval I of E which is a fundamental subset of E is called a *fundamental subinterval* of E. Note that a subinterval of E is an elementary subset of E while a fundamental subinterval of E is clearly a fundamental subset of E. We remark that it is essential to define these terms distinctively because in general the difference of two elementary sets is not necessarily an elementary set.

A set $D = \{(I_i, x_i) : i = 1, 2, \ldots, n\}$ of interval-point pairs is called a *partial division* of E if $I_1, I_2, \ldots, I_n$ are mutually disjoint subintervals of E such that the set

$$E \setminus \bigcup_{i=1}^{n} I_i$$

is either empty or an elementary subset of E, and for each i, we have $x_i \in \overline{I_i}$. We call the corresponding set of intervals $P = \{I_i : i = 1, 2, \ldots, n\}$ a *partial partition* of E. For each i, we call I_i a *component interval* of D and x_i the *associated point* of I_i. A *division* of E is a partial division $\{(I_i, x_i) : i = 1, 2, \ldots, n\}$ such that the union of I_i is E. The corresponding set of intervals I_i is called a *partition* of E.

For brevity, throughout this book a partial division $\{(I_i, x_i) : i = 1, 2, \ldots, n\}$ will often be written as $D = \{(I, x)\}$ in which (I, x) denotes a typical interval-point pair in D.

Some authors call x_i the *tag* of I_i and D a *tagged partition*. Note that given a partition P there are infinitely many ways of forming a division D by choosing different points in E to be associated points, or tags, of the component intervals. Also note that a partial division of E is essentially a subset of a division of E.

Let $G : \overline{E} \to \mathcal{T}_1$ be a function such that for every $x \in \overline{E}$, we have $x \in G(x) \in \mathcal{T}_1$. We call G a *gauge* on E. If the topology $\mathcal{T}$ of X is induced by a metric d, then $\mathcal{T}_1$ is the set of all d-open balls and a gauge G on E could be given by $G(x) = B(x, \delta(x))$ for each $x \in \overline{E}$, and for some $\delta(x) > 0$. Note that a gauge on E has to be defined on $\overline{E}$ and not just E because for each interval-point pair (I, x) in a partial division of E, the associated point x belongs to $\overline{I}$ and not just I.

Let a gauge G on E be given. An interval-point pair (I, x) is G-*fine* if $I \subseteq G(x)$. A partial division $D = \{(I_i, x_i) : i = 1, 2, \ldots, n\}$ of E is G-fine if (I_i, x_i) is G-fine for each $i = 1, 2, \ldots, n$. Since divisions of E are themselves partial divisions of E, a G-fine division of E is similarly defined.

A partial division D^* of E *refines*, or is a *refinement* of, another partial division D of E if for each $(I, x) \in D^*$, we have $I \subseteq J$ for some $(J, y) \in D$. Likewise, a partial partition P^* of E refines or is a refinement of another partial partition P of E if for each $I \in P^*$, we have $I \subseteq J$ for some $J \in P$.

A gauge G_1 is said to be *finer* than a gauge G_2 on E if for every $x \in \overline{E}$ we have

$$G_1(x) \subseteq G_2(x).$$

Let G_1 and G_2 be two gauges on E. Since $\mathcal{T}_1$ is a basis of $\mathcal{T}$, for each $x \in \overline{E}$, there exists $G(x) \in \mathcal{T}_1$ such that

$$G(x) \subseteq G_1(x) \cap G_2(x).$$

We can then define a gauge G on E which is finer than both G_1 and G_2. Consequently, if D is a G-fine division of E, then D is both G_1-fine and G_2-fine.

1.2   Existence of a Division and the *H*-Integral

In this section, we shall give a constructive proof that given a gauge G on E, a G-fine division of E exists. Since E is a finite union of mutually disjoint intervals, it suffices to prove that a

G-fine division of an interval exists. The assertion that a G-fine division exists for any given gauge G is often referred to as the Cousin's lemma.

The conditions we have imposed so far are not sufficient to ensure the existence of a G-fine division in the general setting. We need an additional condition which we will describe explicitly in Remark 1.1. For simplicity of presentation we shall prove the existence of a division for the case when $\mathcal{T}$ is a locally compact metric topology and indicate subsequently how the result in the general setting can be proved.

Consider the case when the locally compact Hausdorff topology $\mathcal{T}$ of X is induced by a metric d and $\mathcal{T}_1$ is the set of all d-open balls. For each $Y \subseteq X$, the *diameter* of Y is given by

$$\mathrm{diam}(Y) = \sup\{d(x,y) : x, y \in Y\}.$$

Recall that by definition, if Y is compact, then every open cover of Y has a finite subcover. We will need this concept in proving the following theorem.

Theorem 1.1 (Cousin's Lemma) *Given a gauge G on an interval I, a G-fine division of I exists.*

Proof.   We shall prove this result in two parts.

(1) Let us construct a finite collection of open balls which covers $\overline{I}$ such that the centre of each ball lies outside all other open balls. By Condition $(*)$ given on page 5, we can choose an open set U such that $\overline{I} \subseteq U$ and $\mathrm{diam}(U) < +\infty$. Let G_1 be a gauge on I which is finer than G such that for each $x \in \overline{I}$, we have $G_1(x) \subseteq U$ and

$$x \in G_1(x) \subseteq \overline{G_1(x)} \subseteq G(x).$$

This is to ensure that we can define a G-fine division of I subsequently. For each $x \in \overline{I}$, let

$$G_1(x) = B(x, \delta(x))$$

for some $\delta(x) > 0$ and define

$$a_1 = \sup\{\delta(x) : x \in \overline{I}\}.$$

Then $0 < a_1 < \text{diam}(U)$ and so a_1 is finite. Let

$$B_1^{(1)} = B\left(x_1^{(1)}, \frac{a_1}{2}\right)$$

for some $x_1^{(1)} \in \overline{I}$ such that $\delta(x_1^{(1)}) \geq \frac{a_1}{2}$. Note that by the definition of a_1, the open ball $B_1^{(1)}$ exists. Next, choose a point $x_2^{(1)} \in \overline{I}$ such that $\delta(x_2^{(1)}) \geq \frac{a_1}{2}$ and

$$B\left(x_2^{(1)}, \frac{a_1}{2}\right) \bigcap B_1^{(1)} = \emptyset,$$

and define

$$B_2^{(1)} = B\left(x_2^{(1)}, \frac{a_1}{2}\right).$$

Then choose a point $x_3^{(1)} \in \overline{I}$ such that $\delta(x_3^{(1)}) \geq \frac{a_1}{2}$ and

$$B\left(x_3^{(1)}, \frac{a_1}{2}\right) \bigcap \left(B_1^{(1)} \bigcup B_2^{(1)}\right) = \emptyset,$$

and define

$$B_3^{(1)} = B\left(x_3^{(1)}, \frac{a_1}{2}\right).$$

Continue this process inductively to obtain $B_4^{(1)}, B_5^{(1)}, \ldots, B_{n(1)}^{(1)}$ such that for $j = 1, 2, \ldots, n(1)$, we have $B_j^{(1)} = B\left(x_j^{(1)}, \frac{a_1}{2}\right)$ where $x_j^{(1)} \in \overline{I}$, $\delta(x_j^{(1)}) \geq \frac{a_1}{2}$ and

$$B_j^{(1)} \bigcap \left(\bigcup_{i=1}^{j-1} B_i^{(1)}\right) = \emptyset,$$

and there is not another point $x_{n(1)+1}^{(1)} \in \overline{I}$ satisfying both $\delta(x_{n(1)+1}^{(1)}) \geq \frac{a_1}{2}$ and

$$B\left(x_{n(1)+1}^{(1)}, \frac{a_1}{2}\right) \bigcap \left(\bigcup_{i=1}^{n(1)} B_i^{(1)}\right) = \emptyset.$$

Note that since $\operatorname{diam}(U) < +\infty$, there can only be finitely many such open balls $B_i^{(1)}$. Let $\mathcal{B}_1 = \left\{ B_1^{(1)}, B_2^{(1)}, \ldots, B_{n(1)}^{(1)} \right\}$ and define

$$a_2 = \sup\left\{ \delta(x) : x \in \overline{I} \setminus \bigcup_{i=1}^{n(1)} B_i^{(1)} \right\}.$$

Repeat the above process, that is, let

$$B_1^{(2)} = B\left(x_1^{(2)}, \frac{a_2}{2} \right)$$

for some $x_1^{(2)} \in \overline{I} \setminus \bigcup_{i=1}^{n(1)} B_i^{(1)}$ such that $\delta(x_1^{(2)}) \geq \frac{a_2}{2}$. Note that for $i = 1, 2, \ldots, n(1)$, we have $x_1^{(2)} \notin B_i^{(1)}$ and $x_i^{(1)} \notin B_1^{(2)}$. Continue this process inductively like before, we obtain $\mathcal{B}_2 = \left\{ B_1^{(2)}, B_2^{(2)}, \ldots, B_{n(2)}^{(2)} \right\}$ such that for $j = 1, 2, \ldots, n(2)$, we have $B_j^{(2)} = B\left(x_j^{(2)}, \frac{a_2}{2} \right)$ where $x_j^{(2)} \in \overline{I} \setminus \bigcup_{i=1}^{n(1)} B_i^{(1)}$, $\delta(x_j^{(2)}) \geq \frac{a_2}{2}$ and

$$B_j^{(2)} \bigcap \left(\bigcup_{i=1}^{j-1} B_i^{(2)} \right) = \emptyset,$$

and there is not another point $x_{n(2)+1}^{(2)} \in \overline{I} \setminus \bigcup_{i=1}^{n(1)} B_i^{(1)}$ satisfying both $\delta(x_{n(2)+1}^{(2)}) \geq \frac{a_2}{2}$ and

$$B\left(x_{n(2)+1}^{(2)}, \frac{a_2}{2} \right) \bigcap \left(\bigcup_{i=1}^{n(2)} B_i^{(2)} \right) = \emptyset.$$

Note that by our construction, for $i = 1, 2, \ldots, n(1)$ and $j = 1, 2, \ldots, n(2)$, we have $x_i^{(1)} \notin B_j^{(2)}$ and $x_j^{(2)} \notin B_i^{(1)}$. Continue this process of obtaining inductively a finite collection of open balls with radius $\dfrac{a_k}{2}$ for $k = 2, 3, \ldots$, where

$$a_k = \sup \left\{ \delta(x) : x \in \overline{I} \setminus \bigcup_{j=1}^{k-1} \bigcup_{i=1}^{n(j)} B_i^{(j)} \right\},$$

we can proceed to define, for $k = 3, 4, \ldots$, the collection $\mathcal{B}_k = \left\{ B_1^{(k)}, B_2^{(k)}, \ldots, B_{n(k)}^{(k)} \right\}$ such that for $p = 1, 2, \ldots, n(k)$, we have $B_p^{(k)} = B\left(x_p^{(k)}, \dfrac{a_k}{2} \right)$ where

$$x_p^{(k)} \in \overline{I} \setminus \bigcup_{j=1}^{k-1} \bigcup_{i=1}^{n(j)} B_i^{(j)},$$

with $\delta(x_p^{(k)}) \geq \dfrac{a_k}{2}$ and

$$B_p^{(k)} \bigcap \left(\bigcup_{i=1}^{p-1} B_i^{(k)} \right) = \emptyset,$$

and there is not another point $x_{n(k)+1}^{(k)} \in \overline{I} \setminus \bigcup_{j=1}^{k-1} \bigcup_{i=1}^{n(j)} B_i^{(j)}$ satisfying both $\delta(x_{n(k)+1}^{(k)}) \geq \dfrac{a_k}{2}$ and

$$B\left(x_{n(k)+1}^{(k)}, \dfrac{a_k}{2} \right) \bigcap \left(\bigcup_{i=1}^{n(k)} B_i^{(k)} \right) = \emptyset.$$

Note that since $0 \leq a_{k+1} \leq a_k$ for all $k = 1, 2, \ldots$, the sequence $\{a_k\}_{k=1}^{\infty}$ is monotone decreasing and bounded below. It follows that $\{a_k\}_{k=1}^{\infty}$ is convergent and we let

$$\lim_{k \to \infty} a_k = a_0.$$

Clearly, $a_k \geq a_0 \geq 0$ for all $k = 1, 2, \ldots$. We shall show that $a_0 = 0$ by contradiction. Suppose $a_0 > 0$ and consider the set

$$\mathcal{B} = \left\{ B\left(x, \dfrac{a_0}{4} \right) : x \in \overline{I} \right\}$$

which is obviously an open cover of $\overline{I}$. Now note that

$$\bigcup_{k=1}^{\infty} \mathcal{B}_k = \left\{ B_i^{(k)} : i = 1, 2, \ldots, n(k) \text{ where } k = 1, 2, \ldots \right\}$$

is a countable collection of d-open balls whose centres are in $\overline{I}$. By our construction, if $p \neq q$, then $x_i^{(p)} \notin B_j^{(q)}$ and $x_j^{(q)} \notin B_i^{(p)}$ for all $i = 1, 2, \ldots, n(p)$ and $j = 1, 2, \ldots, n(q)$. Hence, the distance between $x_i^{(p)}$ and $x_j^{(q)}$ is at least $\max \left(\dfrac{a_p}{2}, \dfrac{a_q}{2} \right)$. On the other hand, for each $k = 1, 2, \ldots$, the distance between $x_i^{(k)}$ and $x_j^{(k)}$ is at least a_k if $i \neq j$. Thus, the distance between the centres of any two open balls in $\bigcup_{k=1}^{\infty} \mathcal{B}_k$ is more than $\dfrac{a_0}{2}$. As a result, any open ball with radius $\dfrac{a_0}{4}$ can contain at most one of the centres $x_i^{(j)}$. It follows that $\mathcal{B}$ does not have a finite subcover since any finite number of open balls in $\mathcal{B}$ can contain at most a finite number of the centres $x_i^{(j)}$ and thus can not contain $\overline{I}$. However, this contradicts the fact that $\overline{I}$ being compact, every open cover of $\overline{I}$ has a finite subcover. Therefore, $a_0 = 0$. Next, we shall prove that

$$\bigcup_{k=1}^{\infty} \bigcup_{i=1}^{n(k)} B_i^{(k)} \supseteq \overline{I}.$$

Suppose there is $x_0 \in \overline{I} \backslash \bigcup_{k=1}^{\infty} \bigcup_{i=1}^{n(k)} B_i^{(k)}$. Then

$$x_0 \in \overline{I} \backslash \bigcup_{k=1}^{r-1} \bigcup_{i=1}^{n(k)} B_i^{(k)}$$

for $r = 2, 3, \ldots$, which implies that $0 \leq \delta(x_0) \leq a_r$ for all $r = 2, 3, \ldots$. Since $\lim\limits_{r \to \infty} a_r = a_0 = 0$, it follows that $\delta(x_0) = 0$, which is a contradiction because δ is a positive function. Hence

$\displaystyle\bigcup_{k=1}^{\infty} \mathcal{B}_k$ is an open cover of $\overline{I}$. By the compactness of $\overline{I}$, there

exists $\{B_1, B_2, \ldots, B_n\} \subseteq \displaystyle\bigcup_{k=1}^{\infty} \mathcal{B}_k$ such that $\overline{I} \subseteq \displaystyle\bigcup_{i=1}^{n} B_i$. We let

the centre of B_i be y_i for each $i = 1, 2, \ldots, n$.

(2) We shall construct a division of I and show that it is G-fine. First, we let

$$I_1 = I \cap \overline{B_1},$$

$$I_i = I \cap \bigcap_{k=1}^{i-1} (\overline{B_i} \setminus \overline{B_k}), \quad i = 2, 3, \ldots, n.$$

Note that $y_k \notin \overline{B_i}$ if $k \neq i$, and thus $B_k \not\subseteq B_i$ if $k \neq i$. Hence, each I_i is a generalised interval by our definition. Furthermore, note that $y_1 \in \overline{B_1}$ and for each $i = 2, 3, \ldots, n$, the point y_i belongs to $\displaystyle\bigcap_{k=1}^{i-1} (\overline{B_i} \setminus \overline{B_k})$. Since each y_i belongs to $\overline{I}$, we conclude that $y_i \in \overline{I_i}$ for each $i = 1, 2, \ldots, n$. We shall verify that $D = \{(I_i, y_i) : i = 1, 2, \ldots, n\}$ is a G-fine division of I. By our construction, it is clear that

$$\bigcup_{i=1}^{n} I_i = I \cap \left(\bigcup_{i=1}^{n} \overline{B_i} \right).$$

Since

$$I \subseteq \overline{I} \subseteq \bigcup_{i=1}^{n} B_i \subseteq \bigcup_{i=1}^{n} \overline{B_i},$$

it follows that $\displaystyle\bigcup_{i=1}^{n} I_i = I$. It is easy to verify that the intervals I_i are mutually disjoint. Indeed, a point in I_i is in $\overline{B_i}$ and not in any $\overline{B_k}$ if $k \neq i$ and thus cannot belong to I_k if $k \neq i$. Finally, we recall that by our construction, the radius of each B_i is $\dfrac{a_{k(i)}}{2}$, for some positive integer $k(i)$, and $\delta(y_i) \geq \dfrac{a_{k(i)}}{2}$. Hence for each $i = 1, 2, \ldots, n$,

$$I_i \subseteq \overline{B_i} = \overline{B(y_i, \frac{a_{k(i)}}{2})} \subseteq \overline{B(y_i, \delta(y_i))} = \overline{G_1(y_i)} \subseteq G(y_i).$$

Therefore, the division D of I is G-fine.

$\square$

Remark 1.1 *Observe that in the above proof, part (2) can be made metric-independent. As for part (1), the essence is to construct a finite collection of open balls which covers $\overline{I}$ such that the centre of each ball lies outside all other open balls. So to prove the result in general we need to impose the following condition.*

Let a gauge G on an elementary set E be given. Then there exist $x_1, x_2, \ldots, x_n \in \overline{E}$ and gauges $G_1, G_2, \ldots, G_n$ on E such that $G_i(x_i) \subseteq G(x_i)$ for $i = 1, 2, \ldots, n$ while $x_i \notin G_j(x_j)$ if $i \neq j$, and

$$\bigcup_{i=1}^{n} G_i(x_i) \supseteq \overline{E}.$$

Note that if X is a metric space, then the above condition is satisfied as is seen in the proof of Theorem 1.1.

We are aware that the metrizability of $\mathcal{T}$ plays a vital role in the proof of Theorem 1.1. However, as far as possible, we shall keep the setting in this book at a more general level to allow any possible extensions beyond metric spaces.

Our next task is to define the H-integral and prove its uniqueness. We shall also show that the H-integral includes as a special case the well-known Henstock–Kurzweil integral defined on the real line.

Note that since intervals are measurable subsets of E, which we assume to be of finite measure, $\iota(I)$ is well-defined for all subintervals I of E. Also, with Theorem 1.1, it is now meaningful to define Riemann sums. For brevity and where there is no ambiguity, $D = \{(I, x)\}$ shall denote a finite collection of interval-point pairs (I, x), and the corresponding Riemann sum will be denoted by $(D) \sum f(x)\iota(I)$. All functions f considered in this book are real-valued point functions defined on $\overline{E}$.

Given a partial division $D = \{(I, x)\}$ of E and $Y \subseteq \overline{E}$, we will often use the notations $(D) \sum\limits_{x \in Y}$ and $(D) \sum\limits_{x \notin Y}$ to represent the sums over all $(I, x) \in D$ such that $x \in Y$ and $x \notin Y$ respectively. Note that

$$(D) \sum = (D) \sum_{x \in Y} + (D) \sum_{x \notin Y}.$$

Definition 1.1 Let f be a real-valued function on $\overline{E}$. Then f is said to be H-integrable on E to a real number A if for every $\varepsilon > 0$, there exists a gauge G on E such that for any G-fine division $D = \{(I, x)\}$ of E, we have

$$\left| (D) \sum f(x)\iota(I) - A \right| < \varepsilon \tag{1.1}$$

and we write

$$(H) \int_E f = A.$$

We shall sometimes call the real value A the *integral value* of f on E.

The family of all functions f which are H-integrable on E shall be denoted by $\mathcal{H}(E)$. We shall sometimes write $f \in \mathcal{H}(E)$ to mean f is H-integrable on E. The H-integrability of f on any elementary subset E_1 of E is similarly defined.

It is easy to see that the H-integral is uniquely determined. Suppose that for a given $\varepsilon > 0$, inequality (1.1) in Definition 1.1 is satisfied with $A = A_1$ and $G = G_1$, and with $A = A_2$ and

$G = G_2$ respectively. Define a gauge G_0 on E such that $G_0(x) = G_1(x) \cap G_2(x)$ for each $x \in \overline{E}$. Then G_0 is a gauge which is finer than both G_1 and G_2 such that for any G_0-fine division $D = \{(I, x)\}$ of E, we have, for $j = 1$ and 2,

$$\left| (D) \sum f(x) \iota(I) - A_j \right| < \varepsilon.$$

By Cousin's lemma (Theorem 1.1), a G_0-fine division $D = \{(I, x)\}$ of E exists. Consequently, and by the triangle inequality, we obtain

$$
\begin{aligned}
|A_1 - A_2| &< \left| (D) \sum f(x) \iota(I) - A_1 \right| \\
&\quad + \left| (D) \sum f(x) \iota(I) - A_2 \right| \\
&< 2\varepsilon.
\end{aligned}
$$

The uniqueness of the H-integral then follows from the arbitrariness of ε.

In the following example, we shall verify that continuous functions are H-integrable. A real-valued point function f defined on $\overline{E}$ is *continuous* at $x \in \overline{E}$ if for every $\varepsilon > 0$, there exists $U \in \mathcal{T}_1$ with $x \in U$ such that for every $y \in U \cap \overline{E}$, we have

$$|f(y) - f(x)| < \varepsilon.$$

We say that f is continuous on $\overline{E}$ if f is continuous at all $x \in \overline{E}$.

Example 1.4 Suppose that f is a continuous function on $\overline{E}$. We let $\{\varepsilon_n\}_{n=1}^{\infty}$ be a sequence of positive real numbers which is monotone decreasing to 0 and for each positive integer n, let G_n be a gauge on E such that whenever $x \in G_n(\xi) \cap \overline{E}$, we have

$$|f(x) - f(\xi)| < \varepsilon_n.$$

We may assume G_{n+1} to be finer than G_n for all n. For $i = 1, 2, \ldots,$ we let D_i be a fixed G_i-fine division of E. Now for any positive integers m and n, we let $D_m = \{(I, x)\}$ and $D_n =$

$\{(J, y)\}$. Note that if $I \cap J$ is non-empty and contains t, then $t \in G_m(\xi) \cap \overline{E}$ and $t \in G_n(\xi) \cap \overline{E}$ and thus

$$|f(x) - f(y)| \le |f(x) - f(t)| + |f(t) - f(y)|$$
$$< \varepsilon_m + \varepsilon_n.$$

It follows that

$$\left|(D_m)\sum f(x)\iota(I) - (D_n)\sum f(y)\iota(J)\right| < (\varepsilon_m + \varepsilon_n)\iota(E),$$

and hence

$$A = \lim_{n \to \infty}\left[(D_n)\sum f(x)\iota(I)\right]$$

exists. Now given $\varepsilon > 0$ choose a positive integer N such that $\varepsilon_N < \varepsilon$ and

$$\left|(D_N)\sum f(x)\iota(I) - A\right| < \varepsilon.$$

Therefore, for any G_N-fine division $D = \{(I, x)\}$ of E, we have

$$\left|(D)\sum f(x)\iota(I) - A\right| \le \left|(D)\sum f(x)\iota(I) - (D_N)\sum f(x)\iota(I)\right|$$
$$+ \left|(D_N)\sum f(x)\iota(I) - A\right|$$
$$< 2\varepsilon \times \iota(E) + \varepsilon.$$

It follows that f is H-integrable on E.

We remark that in the above proof, the technique of constructing a Cauchy sequence of Riemann sums is useful when the integral of the function is not given.

In what follows, we shall show that the Henstock–Kurzweil integral defined on the real line (see, for example, [21, page 5]) is a special case of the H-integral.

Example 1.5 Let X be the real line $\mathbb{R}$ and $\mathcal{T}$ be the topology induced by the usual metric. As illustrated in Example 1.1, $\mathcal{T}_1$ is the set of all intervals of the form (u, v) and $\mathcal{I}_1$ is the set of all bounded intervals of the form (u, v), $(u, v]$, $[u, v)$ and $[u, v]$.

Let $E = [a, b]$ which is an elementary set by our definition. A gauge G on E is such that for each $x \in [a, b]$, we have

$$G(x) = (x - \delta(x), x + \delta(x))$$

where $\delta(x) > 0$. Other related definitions, in particular that of H-integrability on $[a, b]$ in terms of G-fine divisions can be translated to the real line language in a similar manner. Now let us define the Henstock–Kurzweil integral. If $\delta(\xi) > 0$ for $\xi \in [a, b]$, then it is well known that the Heine–Borel theorem yields $a = x_0 < x_1 < \cdots < x_n = b$ and $\xi_1, \xi_2, \ldots, \xi_n \in [a, b]$ such that

$$\xi_i \in [x_{i-1}, x_i] \subseteq (\xi_i - \delta(\xi_i), \xi_i + \delta(\xi_i))$$

for $i = 1, 2, \ldots, n$. We say that a δ-fine division

$$D = \{([x_{i-1}, x_i], \xi_i) : i = 1, 2, \ldots, n\}$$

of $[a, b]$ exists. A real-valued function f on $\overline{E}$ is said to be Henstock–Kurzweil integrable to a real number A on $[a, b]$ if for every $\varepsilon > 0$ there is a function $\delta(\xi) > 0$ such that for every δ-fine division D as described above, we have

$$\left| \sum_{i=1}^{n} f(\xi_i)(x_i - x_{i-1}) - A \right| < \varepsilon.$$

Note that each gauge G gives rise to a positive function δ and vice versa. It is then readily seen from the definitions of H-integrability and Henstock–Kurzweil integrability that the H-integral and the Henstock–Kurzweil integral are equivalent. Therefore, the H-integral includes the Henstock–Kurzweil integral as a special case.

Remark 1.2 *In [1], the intervals are defined axiomatically before the integral is defined. Our choice of intervals and setting fit into the model of the former case. It is then easy to check that the H-integral includes the integral defined in [1] and hence includes the Perron and variational integrals. As this is not within the scope of this book, we shall not elaborate here.*

1.3   Simple Properties of the H-Integral

In this section, we shall establish some simple properties of the H-integral. The two main results here are Henstock's lemma and the monotone convergence theorem. The proofs of all the results in this section will be provided to illustrate the various proof techniques which will be used to derive further properties of the H-integral.

Throughout this book, a property is said to hold *almost everywhere* in $\overline{E}$ if it holds everywhere except perhaps in a *set of measure zero*, that is, the property holds for all $x \in \overline{E} \setminus Z$ where $\iota(Z) = 0$. Sometimes we say "for almost all x in $\overline{E}$" in place of "almost everywhere in $\overline{E}$". In this book, we shall consider only real-valued functions which are finite almost everywhere.

The proof of the first proposition we shall present demonstrates one technique of relaxing "everywhere" to "almost everywhere".

Proposition 1.1   *Let E be an elementary set and f be a real-valued function on $\overline{E}$. If $f(x) = 0$ for almost all x in $\overline{E}$, then f is H-integrable to the value 0 on E.*

Proof.   Let $f(x) = 0$ for all $x \in \overline{E} \setminus Z$ where $\iota(Z) = 0$. Note that Z is the union of X_i, $i = 1, 2, \ldots$, where X_i is a subset of Z such that

$$i - 1 \leq |f(x)| < i$$

for $x \in X_i$. Each X_i is also of measure zero since $0 \leq \iota(X_i) \leq \iota(Z) = 0$. Given $\varepsilon > 0$ and for each i, by virtue of Condition $(*)$ given on page 5, we can choose an open set U_i such that

$$\iota(U_i) < \frac{\varepsilon}{2^i \times i}$$

and $X_i \subseteq U_i$. Define a gauge G on E such that $G(x) \subseteq U_i$ for $x \in X_i$, $i = 1, 2, \ldots$, and arbitrarily otherwise. Then for any

G-fine division $D = \{(I, x)\}$ we have

$$\left| (D) \sum f(x)\iota(I) \right| = \left| (D) \sum_{x \in \overline{E} \setminus Z} f(x)\iota(I) + (D) \sum_{x \in Z} f(x)\iota(I) \right|$$

$$\leq \sum_{i=1}^{\infty} \sum_{x \in X_i} |f(x)|\iota(I)$$

$$< \sum_{i=1}^{\infty} \left(i \times \frac{\varepsilon}{2^i \times i} \right)$$

$$= \varepsilon.$$

This completes the proof.

$\square$

Given real-valued functions f and g on E and $\alpha \in \mathbb{R}$, the functions $f + g$ and αf are given by

$$(f + g)(x) = f(x) + g(x)$$

and

$$(\alpha f)(x) = \alpha \times f(x)$$

for each $x \in \overline{E}$.

In the next proposition we shall state several basic properties of the H-integral. Though these properties are either direct consequences of Definition 1.1 or can be proved easily, we shall still provide the proofs here for the reader's convenience.

Proposition 1.2 *Let E be an elementary set and let $\mathcal{H}(E)$ be the set of all H-integrable functions on E.*

(i) *If $f, g \in \mathcal{H}(E)$ and $\alpha \in \mathbb{R}$, then $f + g, \alpha f \in \mathcal{H}(E)$, that is, $\mathcal{H}(E)$ is closed under addition and scalar multiplication. Furthermore,*

$$(H) \int_E (f + g) = (H) \int_E f + (H) \int_E g$$

and

$$(H) \int_E (\alpha f) = \alpha \times (H) \int_E f.$$

(ii) *If c is a real number and $f(x) = c$ for almost all x in $\overline{E}$, then $f \in \mathcal{H}(E)$ and*

$$(H) \int_E f = c \times \iota(E).$$

(iii) *If $f \in \mathcal{H}(E)$ and $f(x) \geq 0$ for almost all x in $\overline{E}$, then*

$$(H) \int_E f \geq 0.$$

(iv) *If $f, g \in \mathcal{H}(E)$ and $f(x) \leq g(x)$ for almost all x in $\overline{E}$, then*

$$(H) \int_E f \leq (H) \int_E g.$$

(v) *If $f \in \mathcal{H}(E)$ and $f(x) \geq 0$ for almost all x in $\overline{E}$ and $(H) \int_E f = 0$, then*

$$(H) \int_E g = 0$$

for each $g \in \mathcal{H}(E)$ such that $|g(x)| \leq f(x)$ for almost all x in $\overline{E}$.

Proof. (i) Let A and B be the integral values of f and g on E and let $\varepsilon > 0$ be given. There exist gauges G_1 and G_2 on E such that for any G_1-fine division $D = \{(I, x)\}$ of E, we have

$$\left| (D) \sum f(x)\iota(I) - A \right| < \min\left(\frac{\varepsilon}{2}, \frac{\varepsilon}{|\alpha|} \right)$$

and for any G_2-fine division $D^* = \{(J, y)\}$ of E, we have

$$\left| (D^*) \sum g(y)\iota(J) - B \right| < \frac{\varepsilon}{2}.$$

For each $x \in \overline{E}$, define $G(x) = G_1(x) \cap G_2(x)$. Then G is a gauge on E which is finer than both G_1 and G_2. Thus, if $D = \{(I, x)\}$ is a G-fine division of E, we have

$$\left| (D) \sum (f + g)(x)\iota(I) - (A + B) \right|$$
$$\leq \left| (D) \sum f(x)\iota(I) - A \right| + \left| (D) \sum g(x)\iota(I) - B \right|$$
$$< \varepsilon.$$

Hence $f + g$ is H-integrable on E to the value $A + B$. On the other hand, for any G_1-fine division $D = \{(I, x)\}$ of E, we have

$$\left| (D) \sum (\alpha f)(x) \iota(I) - \alpha A \right|$$

$$= |\alpha| \left| (D) \sum f(x) \iota(I) - A \right|$$

$$< \varepsilon$$

and so αf is H-integrable on E to the value αA.

(ii) Applying Proposition 1.1 on the function g given by $g(x) = f(x) - c$ for all $x \in \overline{E}$ such that $f(x) = c$, we obtain the result that g is H-integrable to the value 0 on E. Then for any given $\varepsilon > 0$, we choose a gauge G on E such that for any G-fine division $D = \{(I, x)\}$ of E, we have

$$\left| (D) \sum g(x) \iota(I) \right| < \varepsilon.$$

Consequently,

$$\left| (D) \sum f(x) \iota(I) - c \times \iota(E) \right|$$

$$= \left| (D) \sum [g(x) + c] \, \iota(I) - c \times \iota(E) \right|$$

$$= \left| (D) \sum g(x) \iota(I) + c \times (D) \sum \iota(I) - c \times \iota(E) \right|$$

$$< \varepsilon.$$

Hence f is H-integrable on E to the value $c \times \iota(E)$.

(iii) Since the technique used in proving Proposition 1.1 can be employed to handle sets of measure zero, we may assume that $f(x) \geq 0$ for all x in $\overline{E}$. Let $\varepsilon > 0$ be given and choose a gauge G on E such that for any G-fine division $D = \{(I, x)\}$ of E we have

$$\left| (D) \sum f(x) \iota(I) - (H) \int_E f \right| < \varepsilon$$

which, together with the assumption that $f(x) \geq 0$ for all x in $\overline{E}$, yields

$$-\varepsilon \leq (D) \sum f(x) \iota(I) < (H) \int_E f + \varepsilon.$$

Since ε is arbitrary, it follows that $(H)\int_E f \geq 0.$

(iv) We apply assertion (iii) on the function h given by

$$h(x) = g(x) - f(x)$$

for all $x \in \overline{E}$ such that $f(x) \leq g(x)$ and obtain the result that $(H)\int_E h \geq 0$ where

$$(H)\int_E h = (H)\int_E g - (H)\int_E f$$

by virtue of assertion (i). Hence we obtain

$$(H)\int_E f \leq (H)\int_E g$$

as desired.

(v) Since $-f(x) \leq g(x) \leq f(x)$ for almost all x in $\overline{E}$, by assertions (i) and (iv), we have

$$-(H)\int_E f \leq (H)\int_E g \leq (H)\int_E f$$

where $(H)\int_E f = 0$. Therefore $(H)\int_E g = 0$.   □

Next, we note that by applying Proposition 1.1 and Proposition 1.2(i), the following result follows.

Proposition 1.3   *Let E be an elementary set and let $\mathcal{H}(E)$ be the set of all H-integrable functions on E. If the functions f and g on $\overline{E}$ are such that $f(x) = g(x)$ almost everywhere in $\overline{E}$, then $f \in \mathcal{H}(E)$ if and only if $g \in \mathcal{H}(E)$, and in which case,*

$$(H)\int_E f = (H)\int_E g.$$

The above proposition suggests that as far as H-integrability is concerned, sets of measure zero can be ignored.

The next proposition is a Cauchy test of integrability.

Proposition 1.4 (Cauchy Criterion) *Let E be an elementary set and let f be a real-valued function on $\overline{E}$. Then $f \in \mathcal{H}(E)$ if and only if for every $\varepsilon > 0$, there exists a gauge G on E such that for all G-fine divisions $D = \{(I, x)\}$ and $D^* = \{(J, y)\}$ of E, we have*

$$\left| (D) \sum f(x) \iota(I) - (D^*) \sum f(y) \iota(J) \right| < \varepsilon.$$

Proof. The necessity of the Cauchy criterion follows immediately from Definition 1.1. We shall prove the sufficiency. First, by the given hypothesis, for each $n = 1, 2, \ldots$, there exists a gauge G_n on E such that for all G_n-fine divisions $D = \{(I, x)\}$ and $D^* = \{(J, y)\}$ of E, we have

$$\left| (D) \sum f(x) \iota(I) - (D^*) \sum f(y) \iota(J) \right| < \frac{1}{n}.$$

Without losing generality, we may assume that for each n, G_{n+1} is finer than G_n. Next, for each $n = 1, 2, \ldots$, let $D_n = \{(I^{(n)}, x^{(n)})\}$ be a G_n-fine division of E. Then for positive integers $m > n$, we have

$$\left| (D_m) \sum f(x^{(m)}) \iota(I^{(m)}) - (D_n) \sum f(x^{(n)}) \iota(I^{(n)}) \right| < \frac{1}{n}.$$

Thus $\left\{ (D_n) \sum f(x^{(n)}) \iota(I^{(n)}) \right\}_{n=1}^{\infty}$ is a Cauchy sequence of real numbers and so it converges. Hence there exists a real number A such that

$$\lim_{n \to \infty} \left[(D_n) \sum f(x^{(n)}) \iota(I^{(n)}) \right] = A.$$

Now let $\varepsilon > 0$ be given and let M be a positive integer such that $M \geq \dfrac{2}{\varepsilon}$ and

$$\left| (D_M) \sum f(x^{(M)}) \iota(I^{(M)}) - A \right| < \frac{\varepsilon}{2}.$$

Consequently, for any G_M-fine division $D = \{(I, x)\}$ of E, we have

$$\left| (D) \sum f(x)\iota(I) - A \right|$$
$$\leq \left| (D) \sum f(x)\iota(I) - (D_M) \sum f(x^{(M)})\iota(I^{(M)}) \right|$$
$$+ \left| (D_M) \sum f(x^{(M)})\iota(I^{(M)}) - A \right|$$
$$< \frac{1}{M} + \frac{\varepsilon}{2}$$
$$\leq \varepsilon$$

and so the result follows.

$\square$

The proofs of the following two propositions are similar to those of the corresponding real line results. They are provided here for the sake of completeness. We recall that a fundamental subset E_0 of E is a subset of E such that both E_0 and $E \setminus E_0$ are elementary sets.

Proposition 1.5 *Let E be an elementary set and let f be a real-valued function on $\overline{E}$. If $f \in \mathcal{H}(E)$, then $f \in \mathcal{H}(E_0)$ for all fundamental subsets E_0 of E.*

Proof. Given $\varepsilon > 0$, there exists a gauge G on E such that the Cauchy criterion as described in Proposition 1.4 holds. Now let E_0 be a fundamental subset of E. Let $D_1 = \{(I, x)\}$ and $D_2 = \{(J, y)\}$ be G-fine divisions of E_0 and let D_3 be a G-fine division of $E \setminus E_0$. Clearly any component interval of D_1 is disjoint from any component interval of D_3. The same goes with the component intervals of D_2 and those of D_3. Thus $D_1 \cup D_3$ and $D_2 \cup D_3$ are G-fine divisions of E and hence

$$\left| (D_1 \cup D_3) \sum f(x)\iota(I) - (D_2 \cup D_3) \sum f(y)\iota(J) \right| < \varepsilon.$$

Note that

$$(D_k \cup D_3) \sum = (D_k) \sum + (D_3) \sum$$

for $k = 1$ and 2. Consequently,

$$\left| (D_1) \sum f(x)\iota(I) - (D_2) \sum f(y)\iota(J) \right|$$
$$= \left| (D_1 \cup D_3) \sum f(x)\iota(I) - (D_2 \cup D_3) \sum f(y)\iota(J) \right|$$
$$< \varepsilon.$$

The desired result then follows from Proposition 1.4. $\qquad\square$

Remark 1.3 *By definition, if $D = \{(I_i, x_i)\}_{i=1}^n$ is a partial division of E which is not a division of E, then $E \setminus \bigcup_{i=1}^n I_i$ is necessarily an elementary set. Hence for each $i = 1, 2, \ldots, n$, the set $E \setminus I_i$ is an elementary subset of E and thus each I_i is a fundamental subset of E. It follows from the above proposition that if $f \in \mathcal{H}(E)$ then $f \in \mathcal{H}(I)$ for every component interval I of D.*

Proposition 1.6 *Let E be a disjoint union of intervals $J_1, J_2, \ldots, J_m$. If $f \in \mathcal{H}(J_j)$ for each $j = 1, 2, \ldots, m$, then $f \in \mathcal{H}(E)$ and*

$$(H) \int_E f = \sum_{j=1}^m (H) \int_{J_j} f.$$

Proof. Let $\varepsilon > 0$ be given and for each $j = 1, 2, \ldots, m$, let G_j be a gauge on J_j such that for any G_j-fine division $D = \{(I, x)\}$ of J_j, we have

$$\left| (D) \sum f(x)\iota(I) - (H) \int_{J_j} f \right| < \frac{\varepsilon}{m}.$$

For each $x \in \overline{E}$, let $V(x)$ be an open neighbourhood of x such that $V(x) \cap J_j = \emptyset$ if $x \notin \overline{J_j}$ and choose $G(x) \in \mathcal{T}_1$ such that $x \in G(x)$ and

$$G(x) \subseteq V(x) \cap \bigcap_{x \in \overline{J_k}} G_k(x).$$

Note that it is possible that $x \in \bigcap_{j=1}^{m} \partial J_j$ in which case we choose

$$G(x) \subseteq \bigcap_{j=1}^{m} G_j(x).$$

This defines a gauge G on E. Apply Cousin's lemma (Theorem 1.1) to construct a G-fine division $D = \{(I_i, x_i) : i = 1, 2, \ldots, n\}$ of E. By our choice of G, whenever $x_i \notin \overline{J_j}$, we have

$$I_i \cap J_j = \emptyset.$$

For each $j = 1, 2, \ldots, m$, we then let

$$D_j = \left\{ (I_i \cap J_j, x_i) : x_i \in \overline{J_j} \right\}.$$

It is obvious that D_j is a G_j-fine division of J_j for each $j = 1, 2, \ldots, m$. Furthermore,

$$(D) \sum f(x)\iota(I) = \sum_{j=1}^{m} \left[(D_j) \sum f(x)\iota(I) \right].$$

Consequently, we obtain

$$\left| (D) \sum f(x)\iota(I) - \sum_{j=1}^{m} (H) \int_{J_j} f \right|$$

$$\leq \sum_{j=1}^{m} \left| (D_j) \sum f(x)\iota(I) - (H) \int_{J_j} f \right|$$

$$< \varepsilon$$

and the result follows. $\qquad\qquad\qquad\qquad\qquad\qquad\qquad\qquad$ $\square$

The next proposition we shall prove pertains to the H-integrability of the characteristic function. If W is a subset of $\overline{E}$, the *characteristic function* of W, denoted by χ_W, is given by $\chi_W(x) = 1$ if $x \in W$, and 0 otherwise.

Proposition 1.7 *Let Y be a closed subset of $\overline{E}$. Then χ_Y is H-integrable on E to the value $\iota(Y)$.*

Proof. Let $\varepsilon > 0$ be given. By Condition $(*)$, there exists an open set U such that $Y \subseteq U$ and $\iota(U \setminus Y) < \varepsilon$. Define a gauge G on E such that $G(x) \subseteq U$ if $x \in Y$ and

$$G(x) \subseteq \overline{E} \setminus Y$$

if $x \in \overline{E} \setminus Y$. Now let $D = \{(I, x)\}$ be a G-fine division of E. Clearly,

$$(D) \sum_{x \in Y} \chi_Y(x)\iota(I) = (D) \sum_{x \in Y} \iota(I)$$

and

$$(D) \sum_{x \notin Y} \chi_Y(x)\iota(I) = 0.$$

Let E_1 be the union of intervals I such that $(I, x) \in D$ and $x \in Y$. Obviously, $Y \subseteq E_1 \subseteq U$ and thus

$$\left| (D) \sum \chi_Y(x)\iota(I) - \iota(Y) \right| = \left| (D) \sum_{x \in Y} \iota(I) - \iota(Y) \right|$$
$$= \iota(E_1 \setminus Y)$$
$$\leq \iota(U \setminus Y)$$
$$< \varepsilon.$$

This completes the proof. $\qquad\qquad\qquad\qquad\qquad\qquad\qquad\qquad \square$

Next, we shall establish, for the H-integral, the Henstock's lemma (sometimes referred to as the Saks–Henstock lemma) which is an important tool in the Henstock theory of integration.

Proposition 1.8 (Henstock's Lemma) *Let E be an elementary set and let $f \in \mathcal{H}(E)$. For every $\varepsilon > 0$, there exists a gauge G on E such that for any G-fine division $D = \{(I_i, x_i) : i = 1, 2, \ldots, n\}$ of E, we have*

$$\sum_{i=1}^{n} \left| f(x_i)\iota(I_i) - (H) \int_{I_i} f \right| < \varepsilon. \tag{1.2}$$

Proof.   Let $\varepsilon > 0$ be given and let G be a gauge on E such that for any G-fine division $D = \{(I, x)\}$ of E, we have

$$\left| (D) \sum f(x)\iota(I) - (H) \int_E f \right| < \varepsilon.$$

Now let a G-fine division $D = \{(I_i, x_i) : i = 1, 2, \ldots, n\}$ of E be given. By Proposition 1.5 and Remark 1.3, the integral $(H) \int_{I_i} f$ exists for each $i = 1, 2, \ldots, n$. Let $D^{(1)}$ be the collection of all the interval-point pairs in D such that

$$f(x_i)\iota(I_i) - (H) \int_{I_i} f \geq 0$$

and $D^{(2)}$ the collection of the remaining interval-point pairs in D. Next, in view of the H-integrability of f on each I_i, there exist a gauge G^* which is finer than G, and subsequently G^*-fine divisions $D_1, D_2, \ldots, D_n$ on $I_1, I_2, \ldots, I_n$ respectively such that for each $i = 1, 2, \ldots, n$, we have

$$\left| (D_i) \sum f(x)\iota(I) - (H) \int_{I_i} f \right| < \frac{\varepsilon}{n}.$$

Let $D^{(3)}$ and $D^{(4)}$ each be the union of a suitable collection of D_i such that $D^{(1)} \cup D^{(3)}$ and $D^{(2)} \cup D^{(4)}$ are divisions of E. Clearly,

$D^{(1)} \cup D^{(3)}$ and $D^{(2)} \cup D^{(4)}$ are G-fine. Consequently,

$$\sum_{i=1}^{n} \left| f(x_i)\iota(I_i) - (H)\int_{I_i} f \right|$$

$$\leq \left| (D^{(1)}) \sum \left[f(x)\iota(I) - (H)\int_I f \right] \right|$$

$$+ \left| (D^{(2)}) \sum \left[f(x)\iota(I) - (H)\int_I f \right] \right|$$

$$\leq \left| (D^{(1)} \cup D^{(3)}) \sum \left[f(x)\iota(I) - (H)\int_I f \right] \right|$$

$$+ \left| (D^{(3)}) \sum \left[f(x)\iota(I) - (H)\int_I f \right] \right|$$

$$+ \left| (D^{(2)} \cup D^{(4)}) \sum \left[f(x)\iota(I) - (H)\int_I f \right] \right|$$

$$+ \left| (D^{(4)}) \sum \left[f(x)\iota(I) - (H)\int_I f \right] \right|$$

$$< 4\varepsilon$$

and the result follows.

$\square$

Remark 1.4 *Given a G-fine partial division $D = \{(I, x)\}$ of E which is not a division of E, by definition $E_1 := E \backslash \bigcup_{(I,x) \in D} I$ is an elementary set. By Cousin's lemma (Theorem 1.1) a G-fine division D_1 of E_1 exists, and clearly $D \cup D_1$ is a G-fine division of E. As a result, in Henstock's lemma, inequality (1.2) actually holds for all G-fine partial divisions D of E because obviously*

$$(D) \sum \left| f(x)\iota(I) - (H)\int_I f \right|$$

$$\leq (D \cup D_1) \sum \left| f(x)\iota(I) - (H)\int_I f \right|.$$

The last result we shall prove in this chapter is the monotone convergence theorem which is fundamental in the theory of integration.

Theorem 1.2 (Monotone Convergence Theorem) *Let E be an elementary set and let $f_n \in \mathcal{H}(E)$ for $n = 1, 2, \ldots$. If $f_n(x) \to f(x)$ as $n \to \infty$ for almost all x in $\overline{E}$ where $f_1(x) \le f_2(x) \le \cdots$ for almost all x in $\overline{E}$, and $\lim_{n\to\infty} (H) \int_E f_n < +\infty$, then $f \in \mathcal{H}(E)$ and*

$$\lim_{n\to\infty} (H) \int_E f_n = (H) \int_E f.$$

Proof.   For simplicity, we may assume that for all x in $\overline{E}$, we have $f_n(x) \to f(x)$ as $n \to \infty$ and

$$f_1(x) \le f_2(x) \le \cdots.$$

Let $\varepsilon > 0$ be given. By Henstock's lemma (Proposition 1.8) and Remark 1.4, there exist gauges $G_1, G_2, \ldots$ on E such that for each $n = 1, 2, \ldots$, we have

$$(D) \sum \left| f_n(x)\iota(I) - (H) \int_I f_n \right| < \frac{\varepsilon}{2^n} \qquad (1.3)$$

for any G_n-fine partial division $D = \{(I, x)\}$ of E. Let

$$A = \lim_{n\to\infty} (H) \int_E f_n < +\infty.$$

There exists a positive integer N such that for all $n \ge N$,

$$\left| (H) \int_E f_n - A \right| < \varepsilon. \qquad (1.4)$$

Next, since for all $x \in \overline{E}$, we have $f_n(x) \to f(x)$ as $n \to \infty$, for each x in $\overline{E}$, we choose a positive integer $n(x)$ such that $n(x) \ge N$ and

$$|f_{n(x)}(x) - f(x)| < \varepsilon.$$

Then we define a gauge G on E given by

$$G(x) = G_{n(x)}(x)$$

for all $x \in \overline{E}$ and let $D = \{(I_i, x_i) : i = 1, 2, \ldots, p\}$ be a G-fine division of E. We shall prove that

$$\left| \sum_{i=1}^{p} f(x_i) \iota(I_i) - A \right| < \varepsilon[2 + \iota(E)].$$

To this end, we first observe that

$$\left| \sum_{i=1}^{p} f(x_i) \iota(I_i) - \sum_{i=1}^{p} f_{n(x_i)}(x_i) \iota(I_i) \right| \qquad (1.5)$$

$$\leq \sum_{i=1}^{p} \left| f(x_i) - f_{n(x_i)}(x_i) \right| \iota(I_i)$$

$$< \varepsilon \sum_{i=1}^{p} \iota(I_i)$$

$$= \varepsilon \times \iota(E).$$

Since the integers $n(x_i)$ need not be distinct, for convenience, we let

$$\{n(x_1), n(x_2), \ldots, n(x_p)\} = \{k_1, k_2, \ldots, k_q\},$$

where $k_1 < k_2 < \cdots < k_q$. Then $\{1, 2, \ldots, p\}$ is the disjoint union of the sets

$$T_j = \{i : n(x_i) = k_j\},$$

$j = 1, 2, \ldots, q$. Moreover, for each $j = 1, 2, \ldots, q$, the set $\{(I_i, x_i) : i \in T_j\}$ is a G_{k_j}-fine partial division of E. Indeed, if $i \in T_j$, then

$$I_i \subseteq G(x_i) = G_{n(x_i)}(x_i) = G_{k_j}(x_i).$$

Consequently, applying (1.3), we get

$$\left| \sum_{i=1}^{p} f_{n(x_i)}(x_i) \iota(I_i) - \sum_{i=1}^{p} \left[(H) \int_{I_i} f_{n(x_i)} \right] \right| \qquad (1.6)$$

$$\leq \sum_{i=1}^{p} \left| f_{n(x_i)}(x_i) \iota(I_i) - (H) \int_{I_i} f_{n(x_i)} \right|$$

$$= \sum_{j=1}^{q} \sum_{i \in T_j} \left| f_{k_j}(x_i) \iota(I_i) - (H) \int_{I_i} f_{k_j} \right|$$

$$< \sum_{j=1}^{q} \frac{\varepsilon}{2^{k_j}}$$

$$< \varepsilon.$$

Since $N \leq n(x_i) \leq k_q$ for $i = 1, 2, \ldots, p$ and for each n, we have $f_n(x) \leq f_{n+1}(x)$ for all $x \in \overline{E}$, applying inequality (1.4) on f_N and f_{k_j}, and by Propositions 1.5, 1.6 and 1.2(iv), we obtain

$$A - \varepsilon < (H) \int_E f_N$$

$$= \sum_{i=1}^{p} \left[(H) \int_{I_i} f_N \right]$$

$$\leq \sum_{i=1}^{p} \left[(H) \int_{I_i} f_{n(x_i)} \right]$$

$$\leq (H) \int_E f_{k_q}$$

$$< A + \varepsilon$$

which implies that

$$\left| \sum_{i=1}^{p} \left[(H) \int_{I_i} f_{n(x_i)} \right] - A \right| < \varepsilon.$$

Hence, in conjunction with inequalities (1.5) and (1.6) obtained earlier, we have

$$
\left| \sum_{i=1}^{p} f(x_i)\iota(I_i) - A \right|
$$

$$
\leq \left| \sum_{i=1}^{p} f(x_i)\iota(I_i) - \sum_{i=1}^{p} f_{n(x_i)}(x_i)\iota(I_i) \right|
$$

$$
+ \left| \sum_{i=1}^{p} f_{n(x_i)}(x_i)\iota(I_i) - \sum_{i=1}^{p} \left[(H) \int_{I_i} f_{n(x_i)} \right] \right|
$$

$$
+ \left| \sum_{i=1}^{p} \left[(H) \int_{I_i} f_{n(x_i)} \right] - A \right|
$$

$$
< \varepsilon \times \iota(E) + \varepsilon + \varepsilon
$$

$$
= \varepsilon[\iota(E) + 2].
$$

It follows that $f \in \mathcal{H}(E)$ and

$$
(H) \int_E f = A = \lim_{n\to\infty} (H) \int_E f_n.
$$

$\square$

We remark that the above result is consistent with that in [21, Theorem 4.1] which is for the Henstock–Kurzweil integral on the real line.

By our assumption, $\iota(I) = \iota(\overline{I})$ for each interval I. Consequently, $\iota(\partial I) = 0$ for each interval I. Therefore using mutually disjoint or just non-overlapping intervals in a division $D = \{(I, x)\}$ does not affect the value of the corresponding Riemann sum $(D) \sum f(x)\iota(I)$. Two intervals are said to be *non-overlapping* if they have disjoint interiors.

In this book, we choose to consider divisions comprising mutually disjoint intervals, instead of non-overlapping intervals, for simplicity of presentation.

Chapter 2

The Absolute H-Integral and the McShane-Type Integrals

It is well known that on the real line, the absolute Henstock–Kurzweil, the McShane and the Lebesgue integrals are equivalent [21, page 110]. This result has also been generalised to higher dimensional Euclidean spaces. In this chapter, we shall establish the relationship of the H-integral to some McShane-type integrals which parallels the Henstock theory of integration on the real line. Similar results for the H-integral on metric spaces have been proved by the author in [35].

We begin in Section 2.1 with the definition of the M-integral, which is a generalisation of the McShane integral, and then prove that the absolute H-integral and the M-integral are equivalent. The domains of H-integrability and M-integrability will also be extended from elementary sets to measurable sets. This done, in Section 2.2, we shall proceed to establish the equivalence between the M-integral and the Lebesgue integral on measure spaces. This will provide the link between the H-integral and the Davies and Davies–McShane integrals defined by Henstock in [15]. The Davies and Davies–McShane integrals will be introduced in Section 2.3, the main conclusion of which is that for measurable functions, the absolute H-integral, the M-integral, the Davies integral, the Davies–McShane integral and the Lebesgue integral are all equivalent.

2.1   The Absolute H-Integral and the M-Integral

This section aims to prove that a function is M-integrable if and only if it is absolutely H-integrable. We shall generalise to our setting the proof by Wang and Lee in [44] of a similar result which is for the real line.

We reiterate that throughout this book, E is an elementary set of finite measure. A function f is said to be *absolutely H-integrable* on E if f and its absolute value $|f|$ are both H-integrable on E. Given a function f on $\overline{E}$, the function $|f|$ is given by $|f|(x) = |f(x)|$ for all $x \in \overline{E}$.

We first introduce the M-integral. Given a gauge G on E, a finite collection $\{(I_i, x_i) : i = 1, 2, \ldots, n\}$ of interval-point pairs is called a G-*fine McShane partial division* of E if I_i are mutually disjoint subintervals of E such that $I_i \subseteq G(x_i)$ for each $i = 1, 2, \ldots, n$ and where each x_i is in $\overline{E}$ though not necessarily in $\overline{I_i}$. If further to the above conditions, the union of I_i is E, we call D a G-*fine McShane division* of E. Obviously, a G-fine McShane division of E is a G-fine McShane partial division of E. Also, a G-fine partial division of E is a G-fine McShane partial division of E but not conversely.

Note that given a gauge G on E, the collection of all $G(x)$ where $x \in \overline{E}$ is an open cover of $\overline{E}$. Since $\overline{E}$ is compact, there exist $x_1, x_2, \ldots, x_n \in \overline{E}$ such that the union of $G(x_i)$ contains $\overline{E}$. Hence, given a gauge G on E, a G-fine McShane division always exists.

Definition 2.1   Let f be a real-valued function on $\overline{E}$. Then f is said to be M-integrable on E to a real number A if for every $\varepsilon > 0$ there exists a gauge G on E such that for any G-fine McShane division $D = \{(I, x)\}$ of E, we have

$$\left| (D) \sum f(x)\iota(I) - A \right| < \varepsilon$$

and we write $(M) \int_E f = A$.

It is immediate from the definitions of H-integrability and M-integrability that if a function f is M-integrable on E, then it is H-integrable on E, since all G-fine divisions of E are G-fine McShane divisions of E. It is easy to verify that all the results we have proved for the H-integral in Section 1.3 are valid for the M-integral. The corresponding results for the M-integral either follow from those for the H-integral or can be established using the techniques employed in proving the results for the H-integral, though the proofs of some of which may still need some minor adjustments.

In particular, Henstock's lemma (Proposition 1.8) holds for the M-integral where "G-fine partial division" is replaced with "G-fine McShane partial division". Furthermore, a proof very similar to that of Proposition 1.5 will yield the result that if a function f is M-integrable on E, then it is M-integrable on all fundamental subsets E_0 of E.

Let F be a real-valued function defined on the set of all elementary sets. We call such a function an *elementary-set function*. An elementary-set function F is finitely additive (respectively finitely superadditive) over elementary sets if for

$$E_0 = \bigcup_{i=1}^m E_i$$

where the E_i are pairwise disjoint elementary sets, we have

$$F(E_0) = \sum_{i=1}^m F(E_i) \quad \left(\text{respectively } F(E_0) \geq \sum_{i=1}^m F(E_i) \right).$$

The finite additivity and finite superadditivity of F over elementary subsets, fundamental subsets, subintervals, and fundamental subintervals of E are defined in a similar manner.

Given a function f which is H-integrable on E, we can define a function F given by

$$F(E_0) = (H) \int_{E_0} f$$

for each fundamental subset E_0 of E in view of Proposition 1.5. Note that F is an elementary-set function which is finitely additive over fundamental subsets of E by virtue of Proposition 1.6. We call F the *primitive* of f on E. The primitive of an M-integrable function is similarly defined.

The equivalence between M-integrability and absolute H-integrability will be proved in a few theorems. To this end, we need the following definition.

Definition 2.2 An elementary-set function F is said to be AC on E if for every $\varepsilon > 0$ there exists $\eta > 0$ such that for any partial division $D = \{(I, x)\}$ of E satisfying the condition that $(D) \sum \iota(I) < \eta$, we have

$$(D) \sum |F(I)| < \varepsilon. \tag{2.1}$$

Remark 2.1 *In the above definition, if we replace inequality (2.1) with* $\left|(D) \sum F(I)\right| < \varepsilon$*", the resulting definition is equivalent to the original one. Indeed, if the latter inequality holds, then for every partial division $D = \{(I, x)\}$ of E satisfying $(D) \sum \iota(I) < \eta$, we can decompose D into D_1 and D_2, where D_1 contains (I, x) such that $F(I) > 0$ and D_2 otherwise, and obtain*

$$(D) \sum |F(I)| \le \left|(D_1) \sum F(I)\right| + \left|(D_2) \sum F(I)\right|$$
$$< 2\varepsilon.$$

The converse is obvious as

$$\left|(D) \sum F(I)\right| \le (D) \sum |F(I)|.$$

The first theorem we shall prove indicates the relevance of the above concept, which is a type of absolute continuity condition, to M-integrability.

Theorem 2.1 *If a function f is M-integrable on E, then its primitive F is AC on E.*

Proof. Since f is M-integrable on E, by virtue of the Henstock's lemma for the M-integral, for every $\varepsilon > 0$ there is a gauge G on E such that for any G-fine McShane partial division $D = \{(I, x)\}$ of E, we have

$$(D) \sum |F(I) - f(x)\iota(I)| < \frac{\varepsilon}{2}.$$

Let D_0 be one such division and let

$$M = \max\{|f(x)| : (I, x) \in D_0\}.$$

Choose $\eta > 0$ such that $2M\eta < \varepsilon$. Then for any partial division $D = \{(I, x)\}$ of E satisfying $(D) \sum \iota(I) < \eta$, we partition each component interval I of D into subintervals belonging to those in D_0 and denote by D_1 the new collection of intervals with associated points provided by D_0. It is easy to see that D_1 is a G-fine McShane partial division of E. Since F is finitely additive over fundamental subsets of E, it follows that

$$(D) \sum |F(I)| \leq (D_1) \sum |F(I) - f(x)\iota(I)|$$
$$+(D_1) \sum |f(x)|\iota(I)$$
$$< \frac{\varepsilon}{2} + M\eta$$
$$< \varepsilon.$$

Hence F is AC on E. $\qquad\qquad\square$

The following theorem gives a sufficient condition for an H-integrable function to be absolutely H-integrable.

Theorem 2.2 *If a function f is H-integrable on E and bounded on $\overline{E}$, then f is absolutely H-integrable on E.*

Proof. Since f is H-integrable on E, by Henstock's lemma, for every $\varepsilon > 0$ there is a gauge G_1 on E such that for any G_1-fine division $D = \{(I, x)\}$ of E, we have

$$(D) \sum |f(x)\iota(I) - F(I)| < \varepsilon$$

where F is the primitive of f. For every gauge G on E, let

$$A_G = \sup_{D_G} \left((D_G) \sum |F(I)| \right)$$

where the supremum is taken over all G-fine divisions $D_G = \{(I, x)\}$ of E. Next, let

$$A = \inf_G A_G$$

where the infimum is taken over all gauges G on E. Note that since f is bounded, A exists and is finite. Thus, there exists a G_1-fine division D_0 of E such that for any G_1-fine division D of E which refines D_0 we have

$$\left| A - (D) \sum |F(I)| \right| < \varepsilon.$$

Now since X is regular, we can choose a gauge G_2 which is finer than G_1 such that every G_2-fine division D of E can be partitioned into a refinement D' of D_0. For convenience we still write D for D'. Then for any G_2-fine division $D = \{(J, \xi)\}$ of E, we obtain

$$\left| (D) \sum |f(\xi)|\iota(J) - A \right| \le \left| (D) \sum |f(\xi)|\iota(J) - (D) \sum |F(J)| \right|$$
$$+ \left| (D) \sum |F(J)| - A \right|$$
$$< 2\varepsilon.$$

Hence, f is absolutely H-integrable on E. □

Remark 2.2 *In the above proof the boundedness of f is used only to prove that A is finite. Hence Theorem 2.2 holds true as long as A is finite. The latter is true, for example, when the primitive F of f is AC on E.*

The value A defined in the above proof is called the *Henstock variation* of F on E. Note that if f is absolutely H-integrable on E then the Henstock variation can also be given by

$$A = \inf_G \sup_{D_G} \left((D_G) \sum |f(x)| \iota(I) \right).$$

In the next theorem, we will use the fact that H-integrable functions are measurable, the proof of which will be given in Section 3.1. A function f is *measurable* if for any real number c, the set $\{x \in \overline{E} : f(x) > c\}$ is measurable.

Theorem 2.3 *If a function f is H-integrable on E, then f^N is also H-integrable on E where $f^N(x) = f(x)$ when $0 < f(x) \leq N$ and 0 otherwise.*

Proof. For each gauge G on E, let

$$A_G = \sup_{D_G} \left((D_G) \sum f^N(x) \iota(I) \right),$$

where the supremum is taken over all G-fine divisions $D_G = \{(I, x)\}$ of E, and then let

$$A = \inf_G A_G$$

where the infimum is taken over all gauges G on E. Since f^N is bounded, A_G is finite for each gauge G on E and thus so is A. We shall prove that f^N is H-integrable to the value A on E. Given $\bar{\varepsilon} > 0$, we let

$$W = \{x \in \overline{E} : 0 < f(x) \leq N\}$$

and then define

$$W_k = W \cap \{x \in \overline{E} \ : \ (k-1)\varepsilon < f(x) \le k\varepsilon\},$$

where $k = 1, 2, \ldots, p$ and $p\varepsilon \ge N$. Obviously,

$$\bigcup_{k=1}^{p} W_k = W.$$

Since f is H-integrable on E, each W_k is measurable. With Condition $(*)$ given on page 5 we can find open sets U_k and closed sets Y_k such that $U_k \supseteq W_k \supseteq Y_k$ and

$$\iota(U_k \setminus Y_k) < \frac{\varepsilon}{Np}$$

for $k = 1, 2, \ldots, p$. Note that $Y_1, Y_2, \ldots, Y_p$ are pairwise disjoint closed sets. Write

$$Y = \bigcup_{k=1}^{p} Y_k$$

and for each k, let

$$U_k^* = U_k \setminus (Y \setminus Y_k)$$

which is open since U_k is open and $Y \setminus Y_k$ is closed. Clearly $U_k^* \supseteq Y_k$ and

$$\iota(U_k^* \setminus Y_k) < \frac{\varepsilon}{Np}.$$

Define $G_1 : \overline{E} \to \mathcal{T}_1$ such that for $k = 1, 2, \ldots, p$, we have

$$G_1(x) \subseteq U_k^* \quad \text{when } x \in Y_k,$$

and

$$G_1(x) \subseteq U_k \setminus Y \subseteq U_k \setminus Y_k \quad \text{when } x \in W_k \setminus Y_k,$$

and $G_1(x)$ does not intersect Y when $x \notin W$. Then there is a

G_1-fine division $D_0 = \{(I, \zeta)\}$ of E such that

$$\left| A - (D_0) \sum f^N(\zeta) \iota(I) \right| < \varepsilon.$$

Note that, in view of the definition of G_1, each Y_k is covered by those intervals in D_0 whose associated points lie in Y_k. Next, choose a gauge G_2 which is finer than G_1 on E such that every G_2-fine division D of E can be partitioned into a refinement D' of D_0. For convenience we still write D for D'. Now take any G_2-fine division $D = \{(J, \xi)\}$ of E and note that

$$\left| (D) \sum_{\xi \in W \setminus Y} f(\xi) \iota(J) \right| \leq \sum_{k=1}^{p} \left[(D) \sum_{\xi \in W_k \setminus Y_k} |f(\xi)| \iota(J) \right]$$
$$< p \times N \times \frac{\varepsilon}{Np}$$
$$= \varepsilon,$$

and similarly with D replaced by D_0. Also note that for each $k = 1, 2, \ldots, p$, we have

$$|f(\xi) - f(\zeta)| < \varepsilon$$

whenever $\xi, \zeta \in Y_k$. Therefore, we obtain

$$\left| (D) \sum_{\xi \in Y} f(\xi) \iota(J) - (D_0) \sum_{\zeta \in Y} f(\zeta) \iota(I) \right|$$
$$\leq \sum_{k=1}^{p} \left| (D) \sum_{\xi \in Y_k} f(\xi) \iota(J) - (D_0) \sum_{\zeta \in Y_k} f(\zeta) \iota(I) \right|$$
$$\leq \sum_{k=1}^{p} \left\{ \varepsilon \left[(D) \sum_{\xi \in Y_k} \iota(J) \right] + N \left[(D_0) \sum_{\zeta \in Y_k} \iota(I) - (D) \sum_{\xi \in Y_k} \iota(J) \right] \right\}$$
$$< \varepsilon \times \iota(E) + pN \times \frac{\varepsilon}{Np}$$
$$= \varepsilon(\iota(E) + 1).$$

It follows that

$$\left| (D) \sum f^N(\xi)\iota(J) - A \right|$$

$$\leq \left| (D) \sum f^N(\xi)\iota(J) - (D_0) \sum f^N(\zeta)\iota(I) \right|$$

$$+ \left| (D_0) \sum f^N(\zeta)\iota(I) - A \right|$$

$$\leq \left| (D) \sum_{\xi \in Y} f(\xi)\iota(J) - (D_0) \sum_{\zeta \in Y} f(\zeta)\iota(I) \right|$$

$$+ \left| (D) \sum_{\xi \in W \backslash Y} f(\xi)\iota(J) \right|$$

$$+ \left| (D_0) \sum_{\zeta \in W \backslash Y} f(\zeta)\iota(I) \right| + \varepsilon$$

$$< \varepsilon(\iota(E) + 1) + 3\varepsilon$$

$$= \varepsilon(\iota(E) + 4).$$

The proof is complete.

$\square$

Remark 2.3 *Let $W = \{x \in \overline{E} : \alpha < f(x) \leq \beta\}$, where α and β are real numbers such that $\alpha < \beta$. If f is H-integrable on E then the function f_W, given by $f_W(x) = f(x)$ if $x \in W$ and 0 otherwise, can be shown to be H-integrable on E by modifying the proof of Theorem 2.3. Therefore, f_W is absolutely H-integrable on E by Theorem 2.2. We will need this result in the proof of the theorem that follows the next definition.*

The above remark also suggests a way of extending the domains of H-integrability and M-integrability to measurable sets. For every measurable subset W of E, recall that χ_W denotes the characteristic function of W on $\overline{E}$, that is, $\chi_W(x) = 1$ when $x \in W$ and 0 otherwise. Then given a function f on $\overline{E}$, the function $f\chi_W$ is given by

$$f\chi_W(x) = f(x) \times \chi_W(x)$$

for all $x \in \overline{E}$, that is, $f\chi_W(x) = f(x)$ if $x \in W$ and 0 otherwise. For brevity, we will sometimes write f_W for $f\chi_W$.

This leads to the following definition.

Definition 2.3 Let W be a measurable subset of $\overline{E}$. A real-valued function f on $\overline{E}$ is said to be H-integrable on W to a real number A if $f\chi_W$ is H-integrable on E to the number A. If $f\chi_W$ is absolutely H-integrable on E, then we say that f is absolutely H-integrable on W.

We shall use the notation $(H)\int_W f = A$ to mean that the H-integral of f on W is A. If f is absolutely H-integrable on W, then it is also meaningful to write $(H)\int_W |f|$. Note that by definition, if f is H-integrable on W to the value A, then

$$(H)\int_W f = (H)\int_E f\chi_W = A.$$

If F is the primitive of f on E, we can now write $F(W) = A$.

The M-integrability of f on a measurable subset of $\overline{E}$ is similarly defined.

As mentioned previously, in defining H-integrability, we can ignore sets of measure zero. Therefore, it is of no consequence whether W is a subset of E or $\overline{E}$ in the above definition. Indeed, since $\iota(E) = \iota(\overline{E})$, when f is H-integrable on $W \subseteq \overline{E}$ in the sense of the above definition, we have

$$(H)\int_W f = (H)\int_{W \cap E} f.$$

Theorem 2.4 *If a function f is H-integrable on E and its primitive F is AC on E, then for every $\varepsilon > 0$ there exists $\eta > 0$ such that for any measurable subset Y of $\overline{E}$ on which f is absolutely H-integrable with $\iota(Y) < \eta$, we have*

$$(H) \int_Y |f| < \varepsilon.$$

Proof. Let $\varepsilon > 0$ be given and let $\eta > 0$ be such that for any partial division $D = \{(I, x)\}$ of E satisfying $(D) \sum \iota(I) < \eta$, we have

$$(D) \sum |F(I)| < \frac{\varepsilon}{4}.$$

Let Y be a measurable subset of $\overline{E}$ on which f is absolutely H-integrable with $\iota(Y) < \dfrac{\eta}{2}$. By Condition $(*)$ on page 5, we can choose an open set U such that $Y \subseteq U$ and

$$\iota(U \setminus Y) < \frac{\eta}{2}.$$

Next, by choosing a gauge G on E such that $x \in G(x) \subseteq U$ if $x \in Y$, and applying Henstock's lemma on $|f\chi_Y|$, we obtain that for all G-fine divisions $D = \{(I, x)\}$ of E,

$$\left| (D) \sum_{x \in Y} |f(x)| \iota(I) - (H) \int_Y |f| \right| < \frac{\varepsilon}{2}. \qquad (2.2)$$

Now let $D = \{(J, \xi)\}$ be a G-fine division of E and note that

$$(D) \sum_{\xi \in Y} \iota(J) \leq \iota(U)$$
$$= \iota(U \setminus Y) + \iota(Y)$$
$$< \eta.$$

Consequently, we have

$$(D) \sum_{\xi \in Y} |F(J)| < \frac{\varepsilon}{4}.$$

In view of Henstock's lemma we may also assume that

$$(D) \sum_{\xi \in Y} |f(\xi)\iota(J) - F(J)| < \frac{\varepsilon}{4}.$$

It follows that

$$(D) \sum_{\xi \in Y} |f(\xi)|\iota(J) < \frac{\varepsilon}{2}.$$

Finally, by applying the above inequality and (2.2), we obtain

$$(H) \int_Y |f| < \left| (D) \sum_{\xi \in Y} |f(\xi)|\iota(J) - (H) \int_Y |f| \right|$$

$$+ (D) \sum_{\xi \in Y} |f(\xi)|\iota(J)$$

$$< \varepsilon$$

as desired. □

The preceding theorem is essential in proving the following result which gives a sufficient condition for an H-integrable function to be M-integrable.

Theorem 2.5 *If a function f is H-integrable on E and its primitive F is AC on E then f is M-integrable on E.*

Proof. Let $\varepsilon > 0$ be given and let $\eta > 0$ be such that the necessary condition of Theorem 2.4 is satisfied. Write

$$X_k = \left\{ x \in \overline{E} \ : \ (k-1)\varepsilon < f(x) \le k\varepsilon \right\}$$

for $k = 0, \pm 1, \pm 2, \ldots$. Next, for each integer k choose an open set $U_k \supseteq X_k$ such that

$$\iota(U_k \setminus X_k) < \frac{\eta}{(|k|+1)2^{|k|+2}}.$$

In view of Remark 2.3, the function f is absolutely H-integrable on X_k for each k. Write

$$Y = \bigcup_{k=-\infty}^{\infty} (U_k \setminus X_k).$$

Then $\iota(Y) < \eta$ and so

$$(H) \int_Y |f| < \varepsilon.$$

Now define a gauge G on E such that $G(\xi) \subseteq U_k$ when $\xi \in X_k$ for $k = 0, \pm 1, \pm 2, \ldots$. For convenience, write $X_\xi = X_k$ when $\xi \in X_k$ for some integer k and assume that $\eta \leq 1$. Consider a G-fine McShane division $D = \{(J, \xi)\}$ of E. Note that if $\xi \in X_\xi$, then for all $x \in J \cap X_\xi$, we have

$$|f(\xi) - f(x)| < \varepsilon,$$

and if $\xi \notin X_\xi$, then $|f(\xi)| < (|k| + 1)\varepsilon$. Furthermore,

$$\iota(J \setminus X_\xi) \leq \iota(U_k \setminus X_k) < \frac{\eta}{(|k| + 1)2^{|k|+2}}.$$

Consequently, for any G-fine McShane division $D = \{(J, \xi)\}$ of E we obtain

$$\left| (D) \sum f(\xi)\iota(J) - F(E) \right|$$

$$= \left| (D) \sum \left[f(\xi)\iota(J) - (H) \int_J f \right] \right|$$

$$\leq (D) \sum \left[(H) \int_J |f(\xi) - f| \right]$$

$$\leq (D) \sum \left[(H) \int_{J \cap X_\xi} |f(\xi) - f| \right] + (D) \sum \left[(H) \int_{J \setminus X_\xi} |f(\xi)| \right]$$

$$+ (D) \sum \left[(H) \int_{J \setminus X_\xi} |f| \right]$$

$$< (D) \sum [\varepsilon \times \iota(J \cap X_\xi)]$$

$$+ (D) \sum \left[(|k| + 1)\varepsilon \times \frac{\eta}{(|k| + 1)2^{|k|+2}} \right] + (H) \int_Y |f|$$

$$\leq \varepsilon [\iota(E) + 2]$$

and the result follows. $\qquad\qquad\qquad\qquad\qquad\qquad\qquad\qquad\square$

Theorem 2.6 *If a function f is absolutely H-integrable on E, then its primitive F is AC on E.*

Proof. For each $n = 1, 2, \ldots$, let

$$f^n(x) = f(x)$$

when $|f(x)| \leq n$ and 0 otherwise. By Theorem 2.3 and Remark 2.3, the function f^n is H-integrable on E. Since f^n is bounded, by Theorem 2.2, the function $|f^n|$ is also H-integrable on E. Now let $\varepsilon > 0$ be given. It is easy to see that the primitive $|F^n|$ of $|f^n|$ is AC on E. Indeed, for any partial division $D = \{(I, x)\}$ of E with $(D) \sum \iota(I) < \dfrac{\varepsilon}{2n}$, we have

$$(D) \sum |F^n|(I) = (D) \sum \left[(H) \int_I |f^n| \right]$$
$$\leq n \times (D) \sum \iota(I)$$
$$< \frac{\varepsilon}{2}.$$

Next, note that for each $x \in \overline{E}$, we have $|f^n(x)| \to |f(x)|$ as $n \to \infty$ and

$$|f^n(x)| \leq |f^{n+1}(x)| \leq |f(x)|$$

for $n = 1, 2, \ldots$. As is seen in the proof of Theorem 2.3, the real value $\lim\limits_{n \to \infty} (H) \int_E |f^n|$ is equal to the Henstock variation A of $|F^n|$ on E, where $A = \inf\limits_{G} \sup\limits_{D_G} \left((D_G) \sum |F^n(I)| \right)$, which is finite as $|f^n|$ is bounded. By the monotone convergence theorem (Theorem 1.2), there exists a positive integer N such that

$$(H) \int_E |f| - (H) \int_E |f^N| < \frac{\varepsilon}{2}.$$

Finally, choose $\eta < \dfrac{\varepsilon}{2N}$, and note that for any partial division

$D = \{(I, x)\}$ of E with $(D) \sum \iota(I) < \dfrac{\varepsilon}{2N}$, we have

$$(D) \sum |F(I)|$$

$$\leq (D) \sum \left[(H) \int_I |f| \right]$$

$$\leq \left\{ (D) \sum \left[(H) \int_I |f| \right] - (D) \sum \left[(H) \int_I |f^N| \right] \right\}$$

$$+ (D) \sum \left[(H) \int_I |f^N| \right]$$

$$\leq (H) \int_E \left[|f| - |f^N| \right]$$

$$+ (D) \sum |F^N| (I)$$

$$< \varepsilon.$$

Hence F is AC on E.

□

With Theorems 2.5 and 2.6, we have therefore established the following result.

Theorem 2.7   *If a function f is absolutely H-integrable on E, then it is M-integrable on E.*

By Theorem 2.1, if f is M-integrable on E, its primitive F is AC on E. As a result, in view of Remark 2.2, a proof similar to that of Theorem 2.2 in which the Henstock variation A is finite due to F being AC on E will yield the following result.

Theorem 2.8   *If f is M-integrable on E, then so is $|f|$.*

We remark that the above result can also be proved by applying the Cauchy criterion for M-integrability.

Since an M-integrable function on E is H-integrable on E, by Theorem 2.7 and Theorem 2.8, we have obtained the following result as desired.

Theorem 2.9 *Let Y be a measurable subset of $\overline{E}$. A function f on $\overline{E}$ is M-integrable on Y if and only if it is absolutely H-integrable on Y.*

The above result further supports the fact that the H-integral is a generalisation of the Henstock–Kurzweil integral to measure spaces endowed with a locally compact metrizable topology because on the real line a function is Henstock–Kurzweil integrable on $[a, b]$ if and only it is McShane integrable there.

2.2   The H-Integral and the Lebesgue Integral

The main objective of this section is to establish the relationship between the M-integral and the Lebesgue integral. We shall also prove that if $\mathcal{T}$ is a locally compact metric topology then a function which is H-integrable on E is Lebesgue integrable on an interval $J_0 \subseteq E$.

We first define the Lebesgue integral. Though there are many equivalent definitions of the Lebesgue integral, the following version is used for our convenience in proving the desired results. Since the Lebesgue theory of integration is well-known, we shall quote freely any results pertaining to the Lebesgue integral.

Definition 2.4 Let f be a real-valued function on $\overline{E}$ and Y be a measurable subset of $\overline{E}$. Then f is Lebesgue integrable on Y if there are real numbers $c_1, c_2, \ldots$, and measurable subsets $X_1, X_2, \ldots$, of Y such that

$$f = \sum_{i=1}^{\infty} c_i \chi_{X_i} \quad \text{and} \quad \sum_{i=1}^{\infty} |c_i| \iota(X_i) < \infty.$$

The Lebesgue integral of f on Y, denoted by $(L) \displaystyle\int_Y f$, is given by

$$(L) \int_Y f = \sum_{i=1}^{\infty} [c_i \times \iota(X_i)].$$

It is well known that the Lebesgue integral thus defined is *absolutely continuous* [17, Theorem 12.34] in the sense that for every $\varepsilon > 0$, there exists $\eta > 0$ such that for every measurable set $W \subseteq Y$ and whenever $\iota(W) < \eta$, we have

$$(L) \int_W |f| < \varepsilon.$$

The above property will be used later. To avoid any confusion with the AC conditions, we shall refer to this property as the absolute continuity of the Lebesgue integral.

We shall now prove that the M-integral and the Lebesgue integral are equivalent. We will need the fact that a measurable function f which is bounded on a measurable set Y is Lebesgue integrable on Y (for a reference, see [17]).

Theorem 2.10 *Let f be a function on $\overline{E}$ and Y be a measurable subset of $\overline{E}$. Then f is M-integrable on Y if and only if f is Lebesgue integrable on Y.*

Proof. Suppose f is M-integrable on Y. In view of Theorem 2.8, we may assume that f is non-negative on Y. For each $n = 1, 2, \ldots$, we define

$$Y_n = \{x \in Y : 0 \le f(x) \le n\}$$

and let $f_n = f \chi_{Y_n}$. Note that $f_n(x) \to f(x)$ almost everywhere in Y as $n \to \infty$. Now since f is M-integrable, so is f_n by Theorems 2.3 and 2.8. Then since M-integrable functions are measurable, each f_n is a bounded measurable function on Y. Consequently, each f_n is Lebesgue integrable on Y. Finally, since $\{f_n\}$ is a monotone sequence of functions on Y and since

$$\lim_{n \to \infty} (L) \int_Y f_n = (M) \int_Y f < \infty,$$

by Levi's theorem [17, Theorem 12.22], the function f is Lebesgue integrable on Y. Conversely, if f is Lebesgue integrable on Y, then by the absolute continuity of the Lebesgue integral, a proof similar to that of Theorem 2.5 will show that f is M-integrable on Y.

$\square$

The next result we shall prove requires the notion of distance in the structure of X. Therefore we shall again consider the case when $\mathcal{T}$ is a locally compact topology induced by a metric d on X with $\mathcal{T}_1$ denoting the set of all d-open balls. We shall prove that if a function f is H-integrable on E then it is Lebesgue integrable on an interval $J_0 \subseteq E$. The corresponding result for the Henstock–Kurzweil integral on the real line is an old and well-known result (see [41, Theorem 1.4]) which has been generalised to higher dimensional Euclidean spaces in [3]. The proof we shall present here is similar to that in the latter.

We begin with some additional but necessary concepts and results. Note that one version of the well-known Baire category theorem states that a non-empty complete metric space is not a countable union of nowhere dense sets. A set Y is *nowhere dense* in W if for every non-empty open set $U \subseteq W$, there exists a non-empty open set $U^* \subseteq U$ such that $U^* \cap Y = \emptyset$. Hence, if a set S is not nowhere dense in W, then there exists a non-empty open set $U_0 \subseteq W$ such that for all non-empty open set $U \subseteq U_0$, we have $U \cap S \neq \emptyset$. We say that S is *dense* in U_0. For a quick reference of the aforementioned concepts, we refer the reader to [38, page 313–314].

Since $\overline{E}$ is compact, it is a complete subspace of X. It follows that $\overline{E}$ is not a countable union of nowhere dense set. This is an important result which we will use later.

Recall that the *diameter* of a non-empty set Y in X is given by $\operatorname{diam}(Y) = \sup\{d(x,y) : x, y \in Y\}$.

Theorem 2.11   *If a function f is H-integrable on E then there exists an interval $J_0 \subseteq \overline{E}$ such that f is Lebesgue integrable on J_0.*

Proof.   By the H-integrability of f on E, we choose a gauge G on E such that if $D = \{(I, x)\}$ and $D^* = \{(J, y)\}$ are G-fine divisions of E, we have

$$\left| (D) \sum f(x)\iota(I) - (D^*) \sum f(y)\iota(J) \right| < \iota(E).$$

For each $x \in \overline{E}$, we let $G(x) = B(x, \delta(x))$ for some $\delta(x) > 0$. For each $n = 1, 2, \ldots$, let

$$A_n = \left\{ x \in \overline{E} : |f(x)| < n \text{ and } \delta(x) > \frac{1}{n} \right\}.$$

Since $\delta(x) > 0$ and $f(x)$ is finite, we have

$$\bigcup_{n=1}^{\infty} A_n = \overline{E}.$$

By the Baire category theorem, not all A_n are nowhere dense. Therefore there exists a positive integer N and a closed ball $J_0 \subseteq \overline{E}$ such that A_N is dense in J_0. Recall that a closed ball is a generalised interval by definition. We may assume that

$$\text{diam}(J_0) < \frac{1}{N}$$

and a G-fine division of $E \setminus J_0$ exists. We shall prove that f is Lebesgue integrable on J_0. Suppose this is not true. Then $(L) \displaystyle\int_{J_0} |f| = \infty$ and so we can choose a measurable set $W \subseteq J_0$ such that

$$\left| (L) \int_W f \right| > 3N \times \iota(E). \tag{2.3}$$

Now define a function g on J_0 by

$$g(x) = \begin{cases} f(x), & x \in W, \\ 0, & \text{otherwise.} \end{cases}$$

Then g is Lebesgue integrable on J_0 and so is H-integrable on J_0 by Theorem 2.10. Also note that

$$(L) \int_W g = (L) \int_W f.$$

Therefore, there exists a gauge G' on J_0, which is finer than G, such that if $D = \{(I, x)\}$ is a G'-fine division of J_0, then

$$\left| (D) \sum g(x)\iota(I) - (L) \int_W g \right| < N \times \iota(E). \tag{2.4}$$

Let $D_1 = \{(I_i, x_i) : i = 1, 2, \ldots, n\}$ be a fixed G'-fine division of J_0. Choose a G-fine division D_2 of $E \setminus J_0$ and put $D_3 = D_1 \cup D_2$. Clearly D_3 is a G-fine division of E. Next, for each i if $x_i \notin W$ we put $y_i = x_i$. If $x_i \in W$, then using the density of A_N in J_0, we choose an $y_i \in A_N \cap I_i$. Now put $D_4 = \{(I_i, y_i) : i = 1, 2, \ldots, n\}$. Since $\delta(y_i) > \dfrac{1}{N}$ while $\mathrm{diam}(J_0) < \dfrac{1}{N}$, and since both x_i and y_i belong to J_0, for each i and for each $\xi \in I_i$, we have

$$d(\xi, y_i) \leq \mathrm{diam}(J_0) < \frac{1}{N} < \delta(y_i).$$

Hence $I_i \subseteq G(y_i)$ for each i and so D_4 is a G-fine division of J_0. It follows that $D_5 := D_2 \cup D_4$ is a G-fine division of E and thus using the fact that $x_i = y_i$ whenever $x_i \notin W$, we have

$$\iota(E) > \left| (D_3) \sum f(x)\iota(I) - (D_5) \sum f(x)\iota(I) \right|$$

$$= \left| (D_1) \sum f(x_i)\iota(I_i) - (D_4) \sum f(y_i)\iota(I_i) \right|$$

$$= \left| (D_1) \sum_{x_i \in W} f(x_i)\iota(I_i) - (D_4) \sum_{y_i \in W} f(y_i)\iota(I_i) \right|$$

$$\geq \left| (D_1) \sum g(x_i)\iota(I_i) \right| - \left| (D_4) \sum_{y_i \in W} f(y_i)\iota(I_i) \right|.$$

Note that applying (2.3) and (2.4) yields

$$\left| (D_1) \sum g(x_i) \iota(I_i) \right|$$
$$> \left| (L) \int_W g \right| - \left| (D_1) \sum g(x_i) \iota(I_i) - (L) \int_W g \right|$$
$$> 3N \times \iota(E) - N \times \iota(E)$$
$$= 2N \times \iota(E).$$

On the other hand, since $y_i \in A_N$, we have $|f(y_i)| < N$ and so

$$\left| (D_4) \sum_{x_i \in W} f(y_i) \iota(I_i) \right| < N \times \iota(J_0)$$
$$\leq N \times \iota(E).$$

Consequently, we obtain

$$\iota(E) > 2N \times \iota(E) - N \times \iota(E)$$
$$= N \times \iota(E)$$
$$\geq \iota(E),$$

a contradiction proving the desired result.   □

2.3   The Davies Integral and the Davies–McShane Integral

Henstock constructed a division space from an arbitrary non-atomic measure space with a locally compact Hausdorff topology that is compatible with the measure, and defined the Davies–McShane integral in [15]. An *atom* is a measurable set which has positive measure and contains no set of smaller positive measure. A measure which has no atoms is called *non-atomic* or *atomless*. The main objective of this section is to prove that the H-integral includes the Davies–McShane integral. This will be achieved by establishing the equivalence between the Davies–McShane integral and the Lebesgue integral.

Throughout this section, we let a measurable subset Y of $\overline{E}$ be fixed. We shall assume that a singleton set (that is, a set with exactly one element) has measure zero.

We first introduce the Davies integral. We begin with some basic definitions and remarks. Let G be a gauge on E. For a non-empty measurable set $J \subseteq Y$ and $x \in \overline{E}$, the set-point pair (J, x) is *G-fine* if $J \subseteq G(x)$. When J is a generalised interval, the definition coincides with that of an interval-point pair. A *Davies division* of Y is a sequence $\{(J_j, x_j)\}_{j=1}^{\infty}$ of set-point pairs such that J_j are *essentially disjoint* (that is, $\iota(J_j \cap J_k) = 0$ if $j \neq k$) measurable subsets of Y with

$$\iota\left(Y \setminus \bigcup_{j=1}^{\infty} J_j\right) = 0$$

and $\{x_j\}_{j=1}^{\infty} \subseteq \overline{E}$ is a sequence with $x_j \in J_j$ for $j = 1, 2, \ldots$. If (J_j, x_j) is G-fine for $j = 1, 2, \ldots$, we say that $\{(J_j, x_j)\}_{j=1}^{\infty}$ is a G-fine Davies division of Y.

Note that a Davies division is of a somewhat different form from the divisions we have been considering so far. The most distinctive difference is that a Davies division uses a countably infinite collection of essentially disjoint measurable sets instead of a finite collection of pairwise disjoint generalised intervals. Also note that the union of the measurable sets in a Davies division is Y less a set of measure zero instead of Y.

We remark that the definition we present here is as given in [15]. Since sets of measure zero can be ignored in our setting, we can actually replace "essentially disjoint" with "disjoint" and replace "$\iota\left(Y \setminus \bigcup_{j=1}^{\infty} J_j\right) = 0$" with "union Y" in the above definition of a Davies division.

Definition 2.5   The Davies integral of f on a measurable set $Y \subseteq \overline{E}$ is defined as a real number A such that for every $\varepsilon > 0$, there exists a gauge G on E satisfying the condition that for any G-fine Davies division $\{(J_j, x_j)\}_{j=1}^{\infty}$ of Y, we have

$$\left| \sum_{j=1}^{\infty} f(x_j) \iota(J_j) - A \right| < \varepsilon.$$

If the Davies integral A of f on Y exists, we say that f is Davies integrable on Y to the value A and we shall write

$$(DV) \int_Y f = A.$$

Given a gauge G on E, a G-fine Davies division of Y exists. For example, we can first take any $x_1 \in Y$ and define $J_1 = G(x_1) \cap Y$. Subsequently, take $x_2 \in Y$ which does not lie in $G(x_1)$ and define $J_2 = (G(x_2) \setminus G(x_1)) \cap Y$. Continuing this process inductively, we can define, for $j = 2, 3, \ldots,$

$$J_j = \left(G(x_j) \setminus \bigcup_{k=1}^{j-1} G(x_k) \right) \cap Y.$$

Clearly the sets J_j are measurable subsets of Y. Note that though the union of all $G(x_j)$ may not be Y, the difference is at most a set of measure zero. Then $\{(J_j \cup \{x_j\}, x_j)\}_{j=1}^{\infty}$ is a G-fine Davies division of Y since singleton sets are of measure zero.

The above construction shows that the condition $x_j \in J_j$ for $j = 1, 2, \ldots,$ in the definition of a Davies division is superfluous.

To define the Davies–McShane integral, we use Riemann sums involving finite numbers of terms as opposed to countably infinitely many terms in the case of the Davies integral. A *DM-division* D of Y is a finite collection of set-point pairs $\{(J_j, x_j) : j = 1, 2, \ldots, n\}$ where J_j are non-empty disjoint measurable sets with union Y, and each $x_j \in \overline{E}$. Given a gauge G on E, we call $D = \{(J_j, x_j)\}_{j=1}^{n}$ a G-fine *DM-division* of Y if

(J_j, x_j) is G-fine for $j = 1, 2, \ldots, n$. Obviously a G-fine DM-division of Y always exists.

Definition 2.6 The Davies–McShane integral of f on a measurable set $Y \subseteq \overline{E}$ is defined as a real number A such that for every $\varepsilon > 0$, there exists a gauge G on E satisfying the condition that for any G-fine DM-division $D = \{(J_j, x_j)\}_{j=1}^n$ of Y, we have

$$\left| \sum_{j=1}^n f(x_j) \iota(J_j) - A \right| < \varepsilon.$$

If the Davies–McShane integral A of f on Y exists, we say that f is Davies–McShane integrable on Y to the value A and we shall write

$$(DM) \int_Y f = A.$$

As usual, if $D = \{(J_j, x_j)\}_{j=1}^n$ is a G-fine DM-division we may write $D = \{(J, x)\}$ for brevity and $(D) \sum f(x) \iota(J)$ for $\sum_{j=1}^n f(x_j) \iota(J_j)$ where there is no ambiguity.

It is easy to see that both the Davies and the Davies–McShane integrals are uniquely determined. We shall prove that the Davies–McShane integral includes the Davies integral in the following theorem.

Theorem 2.12 *Let Y be a measurable subset of $\overline{E}$ and A be a real number. If a function f is Davies integrable on Y to the value A, then f is Davies–McShane integrable on Y to the same value A.*

Proof. Given any DM-division $D = \{(J, x)\}$ on Y, each set-point pair (J, x) either has $x \in J$ or not. If the former is the case for each $(J, x) \in D$, then D is a Davies division. Otherwise, we see that since

$$\iota(J \cup \{x\}) = \iota(J),$$

the Riemann sum over D is equal to that over D^* where $D^* :=$ $\{(J \cup \{x\}, x)\}$ is a Davies division. It is easy to see from the definitions that if both the Davies and Davies–McShane integrals exist, they have the same value.

$\square$

Note that given two DM-divisions D_1 and D_2 of Y, a division D_3 exists for which the corresponding partition refines the partitions from D_1 and D_2. More precisely, if $(J, x) \in D_3$ there are $(J^{(i)}, x^{(i)}) \in D_i$ such that $J \subseteq J^{(i)}$ for $i = 1, 2$ where x can be either $x^{(1)}$ or $x^{(2)}$ with J running through all non-empty intersections $J^{(1)} \cap J^{(2)}$. Using this idea, we can prove that the Davies–McShane integral is an absolute one. We will need the Cauchy criterion for the Davies–McShane integral which can be easily proved as in Proposition 1.4.

Theorem 2.13 *Let Y be a measurable subset of $\overline{E}$. If a function f is Davies–McShane integrable on Y, then so is $|f|$.*

Proof. Let $\varepsilon > 0$ be given and let G be a gauge on E such that for any two G-fine DM-divisions $D_1 = \{(J^{(1)}, x^{(1)})\}$, $D_2 = \{(J^{(2)}, x^{(2)})\}$ of Y, we have

$$\left| (D_1) \sum f(x^{(1)}) \iota(J^{(1)}) - (D_2) \sum f(x^{(2)}) \iota(J^{(2)}) \right| < \varepsilon. \quad (2.5)$$

With two such divisions D_1 and D_2, let D_3 be as described in the remark preceding the statement of this theorem; that is, $D_3 = \{(J, x)\}$ where $J = J^{(1)} \cap J^{(2)}$ and x is either $x^{(1)}$ or $x^{(2)}$. Let $D_4 = \{(J, y^{(1)})\}$ be such that $f(y^{(1)})$ is the greater of $f(x^{(1)})$ and $f(x^{(2)})$ and let $D_5 = \{(J, y^{(2)})\}$ be such that $f(y^{(2)})$ is the smaller of $f(x^{(1)})$ and $f(x^{(2)})$. Note that D_4 and D_5 are G-fine DM-divisions of Y. Then by applying (2.5) we obtain

$$(D_3) \sum |f(x^{(1)}) - f(x^{(2)})| \iota(J)$$
$$\leq (D_4) \sum f(y^{(1)}) \iota(J) - (D_5) \sum f(y^{(2)}) \iota(J)$$
$$< \varepsilon.$$

Now, since ι is non-negative and countably additive (and thus finitely additive), we have

$$\left|(D_1)\sum |f(x^{(1)})|\iota(J^{(1)}) - (D_2)\sum |f(x^{(2)})|\iota(J^{(2)})\right|$$

$$= \left|(D_3)\sum \left[|f(x^{(1)})| - |f(x^{(2)})|\right]\iota(J)\right|$$

$$\leq (D_3)\sum \left||f(x^{(1)})| - |f(x^{(2)})|\right|\iota(J)$$

$$\leq (D_3)\sum |f(x^{(1)}) - f(x^{(2)})|\iota(J)$$

$$< \varepsilon.$$

Hence, the desired result follows from the Cauchy criterion for Davies–McShane integrability.

$\square$

We shall next prove that the Davies integral includes the Lebesgue integral. The proof we shall present here follows an idea in [5]. The same technique has been employed to prove Theorem 2.5. As this time we are dealing with a countably infinite Riemann sum instead of a finite one, for completeness we shall still provide the proof here.

Theorem 2.14 *Let f be a finite real-valued function on $\overline{E}$ and Y be a measurable subset of $\overline{E}$. Suppose that f is Lebesgue integrable on Y and $(L)\displaystyle\int_Y f = A$. Then f is Davies integrable on Y to the value A.*

Proof. Let $\varepsilon > 0$ be given. By the absolute continuity of the Lebesgue integral, there exists $\eta > 0$ such that for any measurable subset W of Y such that $\iota(W) < \eta$, we have

$$(L)\int_W |f| < \frac{\varepsilon}{3}.$$

For $n = 0, \pm 1, \pm 2, \ldots$, let

$$X_n = \{x \in \overline{E} : (n-1)\alpha < f(x) \leq n\alpha\},$$

where $\alpha = \dfrac{\varepsilon}{3(\eta + \iota(E))}$. Then for each n, we choose an open set $U_n \supseteq X_n$ such that

$$\iota(U_n \setminus X_n) < \frac{\eta}{2^{|n|} \times 3(|n| + 1)}.$$

We define a gauge G on E such that $G(x) \subseteq U_n$ if $x \in X_n$. Now let $I_1, I_2, \ldots$ be essentially disjoint measurable subsets of Y with

$$\iota\left(Y \setminus \bigcup_{i=1}^{\infty} I_i\right) = 0$$

and $x_1, x_2, \ldots$ be points in $\overline{E}$ satisfying $x_i \in I_i \subseteq G(x_i)$ for $i = 1, 2, \ldots$. We shall prove that

$$\left|\sum_{i=1}^{\infty} f(x_i)\iota(I_i) - A\right| < \varepsilon.$$

For each positive integer i, let $n(i)$ be such that $x_i \in X_{n(i)}$. Then $I_i \subseteq U_{n(i)}$ and so

$$I_i \setminus X_{n(i)} \subseteq U_{n(i)} \setminus X_{n(i)}.$$

Consequently, we obtain

$$
\begin{aligned}
\left|\sum_{i=1}^{\infty} f(x_i)\iota(I_i) - A\right| &= \left|\sum_{i=1}^{\infty} (L)\int_{I_i} [f(x_i) - f]\right| \\
&\leq \sum_{i=1}^{\infty} \left[(L)\int_{I_i} |f(x_i) - f|\right] \\
&\leq \sum_{i=1}^{\infty} \left[(L)\int_{I_i \cap X_{n(i)}} |f(x_i) - f|\right] \\
&\quad + \sum_{i=1}^{\infty} \left[(L)\int_{I_i \setminus X_{n(i)}} |f(x_i)|\right] \\
&\quad + \sum_{i=1}^{\infty} \left[(L)\int_{I_i \setminus X_{n(i)}} |f|\right].
\end{aligned}
$$

Let the three terms in the right-hand side of the last inequality above be P, Q and R respectively. We shall show that each sum is less than $\dfrac{\varepsilon}{3}$. First, since for each i and for each $x \in I_i \cap X_{n(i)}$, the real numbers $f(x)$ and $f(x_i)$ both lie in the interval $((n(i) - 1)\alpha, n(i)\alpha]$ whose length is α, we have

$$P \leq \sum_{i=1}^{\infty} \left[(L) \int_{I_i \cap X_{n(i)}} \alpha \right]$$

$$\leq \alpha \sum_{i=1}^{\infty} \iota(I_i)$$

$$= \alpha \times \iota(Y)$$

$$< \frac{\varepsilon}{3}.$$

Next, we collect all those terms in Q, if any, for which $n(i)$ has a given value n, and note that for each such term we have

$$|f(x_i)| \leq (|n| + 1)\alpha.$$

Then by our choice of U_n, we have

$$Q = \sum_{n=-\infty}^{\infty} \sum_{n(i)=n} \left[(L) \int_{I_i \setminus X_n} |f(x_i)| \right]$$

$$\leq \sum_{n=-\infty}^{\infty} \sum_{n(i)=n} [(|n| + 1)\,\alpha \times \iota(I_i \setminus X_n)]$$

$$\leq \sum_{n=-\infty}^{\infty} [(|n| + 1)\alpha \times \iota(U_n \setminus X_n)]$$

$$< \frac{\varepsilon}{3}.$$

Finally, we let

$$W = \bigcup_{i=1}^{\infty} (I_i \setminus X_{n(i)})$$

and note that $R = (L) \int_W |f|$. Also note that

$$\iota(W) = \sum_{n=-\infty}^{\infty} \sum_{n(i)=n} \iota(I_i \setminus X_{n(i)})$$

$$\leq \sum_{n=-\infty}^{\infty} \iota(U_n \setminus X_n)$$

$$< \eta.$$

By our choice of η, it follows that $R < \dfrac{\varepsilon}{3}$. This completes the proof.

$\square$

We have thus proved that the Davies integral includes the Lebesgue integral. Since in Theorem 2.12 we have proved that the Davies–McShane integral includes the Davies integral, it follows that the Davies–McShane integral includes the Lebesgue integral.

It is well known that if Y is measurable, then the characteristic function χ_Y of Y is Lebesgue integrable. With the aforementioned result, we see that χ_Y is also Davies–McShane integrable and

$$\iota(Y) = (DM) \int_Y \chi_Y$$

$$= (L) \int_Y \chi_Y.$$

We shall use the above result to prove a converse of Theorem 2.14 as given in the next theorem.

Theorem 2.15 *If a real-valued function f on $\overline{E}$ is measurable and is Davies–McShane integrable on a measurable subset Y of $\overline{E}$, then f is Lebesgue integrable on Y.*

Proof. First, we note that since

$$f = \frac{|f| + f}{2} - \frac{|f| - f}{2}$$

and the Davies–McShane integral is absolute, we may assume that $f(x) \geq 0$ for all $x \in Y$. For $0 \leq p < q$, we let

$$X(p,q) = \{x \in Y : p \leq f(x) < q\}.$$

Since f is measurable, the characteristic function $\chi_{X(p,q)}$ of $X(p,q)$ is Lebesgue integrable and hence Davies–McShane integrable on E, say to $A(p,q)$. Put $p = 0, \dfrac{1}{n}, \dfrac{2}{n}, \ldots$ and $q = p + \dfrac{1}{n}$, and note that the series $\displaystyle\sum_{j=0}^{\infty} A\left(\dfrac{j}{n}, \dfrac{j+1}{n}\right) \dfrac{j+1}{n}$ is convergent. Also observe that

$$\sum_{j=1}^{\infty} A\left(\frac{j}{n}, \frac{j+1}{n}\right) \frac{j}{n} \leq (DM) \int_Y f \leq \sum_{j=0}^{\infty} A\left(\frac{j}{n}, \frac{j+1}{n}\right) \frac{j+1}{n}.$$

$$(2.6)$$

The difference between the two monotone sums in (2.6) is

$$\sum_{j=0}^{\infty} A\left(\frac{j}{n}, \frac{j+1}{n}\right) \frac{1}{n} = \frac{A(0, \infty)}{n}$$

which tends to zero as $n \to \infty$ since

$$0 \leq A(0, \infty) = \iota(Y) \leq \iota(E) < \infty.$$

Thus, both sums in (2.6) tend to $(DM) \displaystyle\int_Y f$. Therefore f is Lebesgue integrable on Y. $\qquad\square$

With Theorems 2.12, 2.14 and 2.15, we conclude that if f is measurable, then the Davies–McShane integral and the Lebesgue integral of f are equivalent.

The following result follows readily from the above conclusion as well as Theorems 2.9 and 2.10.

Theorem 2.16 *Let Y be a measurable subset of $\overline{E}$. A function f is Davies–McShane integrable on Y if and only if it is absolutely H-integrable on Y.*

With the above theorem, we infer that the H-integral includes the Davies–McShane integral defined in [15]. We have in fact established the result that if f is measurable, then the absolute H-integral, the M-integral, the Davies integral, the Davies–McShane integral, and the Lebesgue integral of f are all equivalent.

Chapter 3

Further Results of the H-Integral

The main objective of this chapter is to establish further results of the H-integral. In Section 3.1, we shall introduce the (LG)-condition [31] and provide a necessary and sufficient condition for H-integrability in terms of this condition. We shall also show that H-integrable functions are measurable and define explicitly a function which is H-integrable but whose absolute value is not to show that the H-integral is indeed a nonabsolute one.

The notions of generalised absolute continuity [22] and equi-integrability, as well as the strong Lusin condition, will be introduced in Section 3.2. Key results involving these concepts will be derived. Several convergence theorems for the H-integral will be proved in Section 3.3. We shall prove the equiintegrability theorem as well as the basic convergence theorem and illustrate how the generalised mean convergence theorem can be proved with the aid of the two aforementioned theorems. The controlled convergence theorem will be proved in a few lemmas and by applying the basic convergence theorem.

Throughout this chapter, we shall assume that every measurable subset Y of $\overline{E}$ has the *Darboux property* (for a reference see [6, page 25]), that is, for every real number α such that $0 \leq \alpha \leq \iota(E)$, there exists a measurable set $W \subseteq \overline{E}$ satisfying $\iota(W) = \alpha$.

71

3.1   A Necessary and Sufficient Condition for H-Integrability

In this section we shall begin with a necessary condition for a function to be H-integrable. This condition will help us prove the measurability of H-integrable functions. A necessary and sufficient condition for H-integrability will be given in terms of the (LG)-condition which we shall define later. We shall then be able to prove that the H-integral is genuinely a nonabsolute one.

The real-line analogue of the following result is due to Lu and Lee [32].

Theorem 3.1   *If a function f is H-integrable on E, then there is a sequence $\{X_k\}$ of closed subsets of $\overline{E}$ such that $X_k \subseteq X_{k+1}$ for all k with $\overline{E}\backslash \bigcup_{k=1}^{\infty} X_k$ being of measure zero, f is Lebesgue integrable on each X_k and*

$$\lim_{k\to\infty}(L)\int_{X_k} f = (H)\int_E f.$$

Proof.   The proof resembles that which shows that a conditionally convergent series can be rearranged to converge to any real number. Consider

$$a_n = (L)\int_{W_n} f \quad \text{and} \quad b_n = (L)\int_{Z_n} f$$

where

$$W_n = \{x \in \overline{E} : n - 1 \le f(x) < n\}$$

and

$$Z_n = \{x \in \overline{E} : -n \le f(x) < -n + 1\}$$

for $n = 1, 2, \ldots$. Let A denote the H-integral of f on E. Obviously, A is the sum of a_n and b_n in some order. Four cases may occur, namely,

(i) $\sum a_n < \infty$ and $\sum b_n > -\infty;$

(ii) $\sum a_n = \infty$ and $\sum b_n > -\infty;$

(iii) $\sum a_n < \infty$ and $\sum b_n = -\infty;$

(iv) $\sum a_n = \infty$ and $\sum b_n = -\infty.$

In the first case, f is Lebesgue integrable and the result follows directly. In the second case, put $f_1(x) = f(x)$ when $x \in \bigcup_{n=1}^{\infty} W_n$ and 0 elsewhere in $\overline{E}$, and $f_2(x) = f(x)$ when $x \in \bigcup_{n=1}^{\infty} Z_n$ and 0 elsewhere in $\overline{E}$. Then, f_2 is Lebesgue integrable and hence H-integrable on E. It follows that $f_1 = f - f_2$ is also H-integrable on E and indeed it is Lebesgue integrable on E. However, this is impossible because $\sum a_n = \infty$. Similarly the third case does not occur. It remains to check the fourth case. First, let $A > 0$ and construct the following two sequences of positive integers. Define $n(1)$ so that

$$\sum_{i=1}^{n(1)-1} a_i \leq A < \sum_{i=1}^{n(1)} a_i.$$

For convenience, we assume that $\sum_{i=1}^{n(1)-1} a_i = 0$ when $n(1) = 1$ and likewise in what follows. Next, define $m(1)$ and $n(2) > n(1)$ so that

$$\sum_{i=1}^{n(1)} a_i + \sum_{i=1}^{m(1)} b_i < A \leq \sum_{i=1}^{n(1)} a_i + \sum_{i=1}^{m(1)-1} b_i$$

and

$$\sum_{i=1}^{n(2)-1} a_i + \sum_{i=1}^{m(1)} b_i \leq A < \sum_{i=1}^{n(2)} a_i + \sum_{i=1}^{m(1)} b_i.$$

Finally, define $m(k)$ and $n(k+1)$ inductively for $k = 2, 3, \ldots,$ so that

$$\sum_{i=1}^{n(k)} a_i + \sum_{i=1}^{m(k)} b_i < A \leq \sum_{i=1}^{n(k)} a_i + \sum_{i=1}^{m(k)-1} b_i,$$

and

$$\sum_{i=1}^{n(k+1)-1} a_i + \sum_{i=1}^{m(k)} b_i \leq A < \sum_{i=1}^{n(k+1)} a_i + \sum_{i=1}^{m(k)} b_i.$$

We shall use the fact that if

$$(L) \int_W f < \alpha < (L) \int_Z f \quad \text{or} \quad (L) \int_W f > \alpha > (L) \int_Z f$$

where W and Z are measurable sets with $W \subseteq Z$, then there exists a measurable set X_0 such that

$$W \subseteq X_0 \subseteq Z \quad \text{and} \quad (L) \int_{X_0} f = \alpha.$$

Note that this is possible because every measurable subset of $\overline{E}$ has the Darboux property. Therefore we may choose measurable sets $X_1, X_2, \ldots$ so that for each $k = 1, 2, \ldots$, we have

$$\left(\bigcup_{i=1}^{n(k)} W_i \right) \cup \left(\bigcup_{i=1}^{m(k)} Z_i \right) \supseteq X_{2k-1} \supseteq \left(\bigcup_{i=1}^{n(k)} W_i \right) \cup \left(\bigcup_{i=1}^{m(k)-1} Z_i \right),$$

$$\left(\bigcup_{i=1}^{n(k+1)-1} W_i \right) \cup \left(\bigcup_{i=1}^{m(k)} Z_i \right) \subseteq X_{2k} \subseteq \left(\bigcup_{i=1}^{n(k+1)} W_i \right) \cup \left(\bigcup_{i=1}^{m(k)} Z_i \right)$$

and

$$(L) \int_{X_k} f = A.$$

Clearly, $X_k \subseteq X_{k+1}$ for all k. If X_k is not closed, then applying Condition $(*)$, we may choose a closed set $Y_k \subseteq X_k$ so that

$$\left| (L) \int_{Y_k} f - A \right| < 2^{-k}.$$

We can verify that all the conditions in the statement of Theorem 3.1 are satisfied with X_k replaced by suitable Y_k if necessary. The case when $A \leq 0$ is similar. The proof is thus complete. $\square$

Corollary 3.1 *If a function f is H-integrable on E, then f is measurable.*

Proof. Let the sequence $\{X_k\}$ of closed sets be as constructed in the proof of Theorem 3.1. For each k, the function $f\chi_{X_k}$ is Lebesgue integrable on E and so is measurable. It is easy to see that

$$\lim_{k\to\infty} f\chi_{X_k}(x) = f(x)$$

almost everywhere in $\overline{E}$. Hence f is measurable. $\qquad\square$

Our main objective in this section is to provide a necessary and sufficient condition for a function to be H-integrable. Theorem 3.1 has shed some light on how such a condition can be formulated. It also leads to the following definition.

Definition 3.1 Let f be a real-valued function defined on $\overline{E}$. A sequence $\{X_i\}$ of measurable sets with union $\overline{E}$ such that f is H-integrable on each X_i is said to be a basic sequence of f on E. The sequence is called monotone increasing if $X_i \subseteq X_{i+1}$ for each i.

The next definition is due to Liu [31] in giving necessary conditions for Henstock–Kurzweil integrability on the real line.

Definition 3.2 Let $\{X_i\}$ be a monotone increasing sequence of measurable subsets of $\overline{E}$. A function f is said to satisfy the (LG)-condition on $\{X_i\}$ if for every $\varepsilon > 0$, there is a positive integer N such that for each $i \geq N$, there is a gauge G_i on E satisfying the condition that

$$\left| (D) \sum_{x \notin X_i} f(x)\iota(I) \right| < \varepsilon \tag{3.1}$$

for all G_i-fine divisions $D = \{(I, x)\}$ of E. A sequence $\{f_n\}$ of functions is said to satisfy the uniform (LG)-condition on $\{X_i\}$ if

(3.1) holds with f replaced by f_n and where the positive integer N and the gauges G_i are independent of n.

The following result is proved by Lee [25] for a Henstock-type integral in the Euclidean space. Here we generalise the proof to our setting. It basically follows from the fact that

$$(D) \sum_{x \notin X_i} = (D) \sum - (D) \sum_{x \in X_i}$$

for each division $D = \{(I, x)\}$ of E.

Theorem 3.2 *Let f be a function on $\overline{E}$ and $\{X_i\}$ be a basic sequence of f on E. Suppose that $\lim_{i \to \infty} (H) \int_{X_i} f = A$. Then f is H-integrable on E to the value A if and only if f satisfies the (LG)-condition on $\{X_i\}$.*

Proof. (Necessity) Let $\varepsilon > 0$ be given. Since f is H-integrable on E, there exists a gauge G on E such that for any G-fine division $D = \{(I, x)\}$ of E, we have

$$\left| (D) \sum f(x) \iota(I) - A \right| < \varepsilon.$$

For convenience, we write

$$A_i = (H) \int_{X_i} f.$$

Then there is a positive integer N such that $|A_i - A| < \varepsilon$ whenever $i \geq N$. Take $i \geq N$ and a gauge G_i which is finer than G such that for any G_i-fine division $D = \{(I, x)\}$ of E we have

$$\left| (D) \sum_{x \in X_i} f(x) \iota(I) - A_i \right| < \varepsilon.$$

It follows that

$$\left| (D) \sum_{x \notin X_i} f(x)\iota(I) \right|$$

$$\leq \left| (D) \sum f(x)\iota(I) - A \right| + |A_i - A|$$

$$+ \left| (D) \sum_{x \in X_i} f(x)\iota(I) - A_i \right|$$

$$< 3\varepsilon.$$

Hence f satisfies the (LG)-condition on $\{X_i\}$.

(Sufficiency) Let $\varepsilon > 0$ and let the integer N be as in Definition 3.2. We may assume that $|A_N - A| < \varepsilon$. Now choose a gauge G on E such that for every G-fine division $D = \{(I, x)\}$ of E, we have

$$\left| (D) \sum_{x \in X_N} f(x)\iota(I) - A_N \right| < \varepsilon.$$

Then for any G-fine division $D = \{(I, x)\}$ of E we obtain

$$\left| (D) \sum f(x)\iota(I) - A \right|$$

$$\leq \left| (D) \sum_{x \in X_N} f(x)\iota(I) - A_N \right| + |A_N - A|$$

$$+ \left| (D) \sum_{x \notin X_N} f(x)\iota(I) \right|$$

$$< 3\varepsilon.$$

Hence f is H-integrable on E to the value A. $\qquad\square$

With Theorems 3.1 and 3.2, we can now provide a necessary and sufficient condition for H-integrability as given in the following theorem.

Theorem 3.3   *A function f is H-integrable on E if and only if f has a basic sequence $\{X_i\}$ on E such that $\lim\limits_{i\to\infty}(H)\displaystyle\int_{X_i} f < +\infty$ and it satisfies the (LG)-condition on $\{X_i\}$.*

Proof.   Suppose f is H-integrable on E and let $\{X_i\}$ be a monotone increasing sequence of closed sets satisfying the conditions of Theorem 3.1. Let

$$Z = \overline{E} \setminus \bigcup_{i=1}^{\infty} X_i$$

be the set of measure zero. Then f is Lebesgue integrable, and hence H-integrable, on the measurable set $X_1 \cup Z$. So $\{X_1 \cup Z, X_2 \cup Z, X_3 \cup Z, \ldots\}$ is a monotone increasing basic sequence of f on E. Since f is H-integrable on E, by Theorem 3.2, f satisfies the (LG)-condition on $\{X_i \cup Z\}_{i=1}^{\infty}$. The converse follows immediately from Theorem 3.2.
$\square$

We shall next prove that the H-integral is nonabsolute by constructing a function which is H-integrable but not absolutely H-integrable.

For convenience we shall assume that the locally compact topology $\mathcal{T}$ is induced by a metric d in X. However, we emphasise that what follows can be made metric-independent.

Example 3.1   First fix an $x_0 \in X$. For $r > 0$, let $B(x_0, r)$ denote the d-open ball centred at x_0 and with radius r as usual. For each positive integer n, we write $B_n = B\left(x_0, \dfrac{1}{n}\right)$ and let

$$Y_n = \overline{B_n} \setminus \overline{B_{n+1}}.$$

Note that for each k, the sets Y_k and Y_{k+1} are disjoint. We then define the function f on $\overline{B_1}$ given by

$$f(x) = \frac{(-1)^{k+1}}{k \times \iota(Y_k)}$$

when $x \in Y_k$ and $f(x_0) = 0$. For each n, let $X_n = \bigcup_{k=1}^{n} Y_k$ and note that

$$\bigcup_{n=1}^{\infty} X_n = \overline{B_1}.$$

Clearly, the function f is Lebesgue integrable and hence H-integrable on X_n for each n. In other words, $\{X_n\}$ is a monotone increasing basic sequence of f on $\overline{B_1}$. Next, for each measurable set $Y \subseteq \overline{B_1}$, we define

$$F(Y) = \sum_{k=1}^{\infty} \frac{(-1)^{k+1}}{k \times \iota(Y_k)} \iota(Y_k \cap Y),$$

if the right-hand side of the equality exists. It is easy to see that

$$F(Y_k) = \frac{(-1)^{k+1}}{k} = (H) \int_{Y_k} f,$$

$$F(\overline{B_1}) = \sum_{k=1}^{\infty} \frac{(-1)^{k+1}}{k}.$$

Now note that for each n and for any measurable set Y such that $\overline{B_{n+1}} \subseteq Y \subseteq \overline{B_{n-1}}$, we have

$$
\begin{aligned}
|F(Y)| &= \left| \sum_{k=1}^{\infty} \frac{(-1)^{k+1}}{k \times \iota(Y_k)} \iota(Y_k \cap Y) \right| \\
&\leq \left| \frac{(-1)^n}{(n-1) \times \iota(Y_{n-1})} \iota(Y_{n-1} \cap Y) \right| \\
&\quad + \left| \frac{(-1)^{n+1}}{n \times \iota(Y_n)} \iota(Y_n \cap Y) \right| \\
&\quad + \left| \sum_{k=n+1}^{\infty} \frac{(-1)^{k+1}}{k} \right| \\
&\leq \frac{1}{n-1} + \frac{1}{n} + \left| \sum_{k=n+1}^{\infty} \frac{(-1)^{k+1}}{k} \right| \\
&= S(n),
\end{aligned}
$$

say, which tends to 0 as $n \to \infty$. Next, let $\varepsilon > 0$ be given and let N be a positive integer such that $S(n) < \varepsilon$ for all $n \geq N$. Then for each $k \geq N$, since f is Lebesgue integrable on Y_k and so is H-integrable there, there exists a gauge G_k on E such that for any G_k-fine partial division $D = \{(I, x)\}$ of E, we have

$$(D) \sum |f\chi_{Y_k}(x)\iota(I) - F(I)| < \frac{\varepsilon}{2^k}.$$

We may assume that for each n, if $x \in \overline{B_n} \setminus \overline{B_{n+1}}$ we have

$$G_k(x) \subseteq \overline{B_{n-1}} \setminus \overline{B_{n+1}},$$

and that if $x \notin \overline{B_n}$ then

$$G_k(x) \cap \overline{B_n} = \emptyset.$$

Now for each $n \geq N$ and for any G_n-fine division $D = \{(J, \xi)\}$ of E, we let D_k denote the collection of interval-point pairs (J, ξ) from D such that $\xi \in Y_k$ for $k = n+1, n+2, \ldots$. Let $Y = \bigcup_{\xi \notin X_n} J$.

By our construction, it is clear that

$$\overline{B_{n+1}} \subseteq Y \subseteq \overline{B_{n-1}}.$$

As a result,

$$\left| (D) \sum_{\xi \notin X_n} f(\xi)\iota(J) \right| \leq \left| (D) \sum_{\xi \notin X_n} [f(\xi)\iota(J) - F(J)] \right|$$

$$+ \left| (D) \sum_{\xi \notin X_n} F(J) \right|$$

$$\leq \sum_{k=n+1}^{\infty} \left| (D_k) \sum [f(\xi)\iota(J) - F(J)] \right|$$

$$+ |F(Y)|$$

$$\leq 2\varepsilon.$$

Hence, f satisfies the (LG)-condition on $\{X_n\}$. It is easy to see that

$$(H) \int_{X_n} f = \sum_{k=1}^{n} \frac{(-1)^{k+1}}{k}.$$

It follows that

$$\lim_{n \to \infty} (H) \int_{X_n} f < \infty.$$

Therefore, by Theorem 3.3, the function f is H-integrable on $\overline{B_1}$. However, since the series $\sum_{k=1}^{\infty} \frac{(-1)^{k+1}}{k}$ is conditionally convergent, the function $|f|$ is not H-integrable on $\overline{B_1}$.

The above example provides a function f which is H-integrable but not absolutely H-integrable. Hence, we have shown that the H-integral is a nonabsolute one.

3.2 Generalised Absolute Continuity and Equiintegrability

It is well known that the primitive of a Lebesgue integrable function is absolutely continuous. On the real line, it has been proved that the primitive of a Henstock–Kurzweil integrable function is generalised absolutely continuous in some restricted sense [21]. We shall see in Chapter 4 that the generalised absolute continuity also plays a vital role in the descriptive definition of the H-integral.

In this section, we shall extend the concept of generalised absolute continuity to the setting of measure spaces endowed with locally compact Hausdorff topologies and introduce the notion of equiintegrability as well as the strong Lusin condition. We shall prove some key results involving these concepts.

We begin with a series of standard definitions relating to generalised absolute continuity.

Definition 3.3 Let $Y \subseteq \overline{E}$ be a measurable set. An elementary-set function F is said to have an absolutely bounded

Riemann sum on Y, or simply said to be $ABRS$ on Y, if there exist a real number c and a gauge G on E such that for any G-fine division $D = \{(I, x)\}$ of E, we have

$$(D) \sum_{x \in Y} |F(I)| \le c.$$

If f is a function on $\overline{E}$ which is absolutely H-integrable on $Y \subseteq \overline{E}$ with primitive F, then it follows from Henstock's lemma that F is $ABRS$ on Y.

Definition 3.4 Let $Y \subseteq \overline{E}$ be a measurable set. An elementary-set function F is said to be $AC_\Delta(Y)$ if F is $ABRS$ on Y and for every $\varepsilon > 0$, there exist a gauge G on E and $\eta > 0$ such that for every G-fine partial division $D = \{(I, x)\}$ of E with $x \in Y$ satisfying the condition that $(D) \sum \iota(I) < \eta$, we have

$$\left| (D) \sum F(I) \right| < \varepsilon. \tag{3.2}$$

If the condition that F is $ABRS$ on Y is omitted in the above definition, we say that F is weakly $AC_\Delta(Y)$. If $\overline{E}$ is the union of closed sets X_i, $i = 1, 2, \ldots$ such that F is $AC_\Delta(X_i)$ (respectively weakly $AC_\Delta(X_i)$) for each i, then F is said to be ACG_Δ (respectively weakly ACG_Δ) on E. If the union of X_i is a proper subset W of $\overline{E}$, then F is said to be ACG_Δ (respectively weakly ACG_Δ) on W.

The above definition differs from that of the AC condition on E in several ways. Other than the obvious fact that the $AC_\Delta(Y)$ condition is defined on any measurable subset Y of $\overline{E}$ while the AC condition is on E, we need the existence of a gauge G in the former in addition to the existence of an η. Furthermore, the $AC_\Delta(Y)$ condition considers partial divisions with associated points in Y instead of full divisions of E. As is pointed out in the remark after Definition 2.2, we may replace (3.2) in the above definition by "$(D) \sum |F(I)| < \varepsilon$".

The next result shows the relation of the concept of ACG_Δ to H-integrability.

Proposition 3.1 *Let f be a function which is H-integrable on E and let F be the primitive of f. Then F is weakly ACG_Δ on $\overline{E} \setminus Z$ where $\iota(Z) = 0$.*

Proof. For each $i = 1, 2, \ldots$, define

$$X_i = \{x \in \overline{E} : |f(x)| < i\}.$$

Note that $\overline{E} = \bigcup_{i=1}^{\infty} X_i$ and $X_i \subseteq X_{i+1}$ for each i. Furthermore, each X_i is measurable. We shall prove that F is weakly $AC_\Delta(X_i)$ for each i. To this end, we let $\varepsilon > 0$ be given. Since f is H-integrable on E, by Henstock's lemma, there exists a gauge G on E such that for any G-fine partial division $D = \{(I, x)\}$ of E, we have

$$(D) \sum |f(x)\iota(I) - F(I)| < \frac{\varepsilon}{2}.$$

Now let $D = \{(I, x)\}$ be a G-fine partial division of E with $x \in X_i$ such that $(D) \sum \iota(I) < \frac{\varepsilon}{2i}$. Then

$$(D) \sum |F(I)| \leq (D) \sum |f(x)\iota(I) - F(I)| + (D) \sum |f(x)| \iota(I)$$
$$< \frac{\varepsilon}{2} + i \times (D) \sum \iota(I)$$
$$< \varepsilon.$$

It follows that F is weakly $AC_\Delta(X_i)$ for each i. Next, by Condition $(*)$, for each i, we can find a closed set W_i such that $W_i \subseteq X_i$ and

$$\iota(X_i \setminus W_i) < \frac{1}{i}.$$

Define

$$Z = \overline{E} \setminus \bigcup_{i=1}^{\infty} W_i.$$

We shall prove that F is weakly ACG_Δ on $\bigcup_{i=1}^{\infty} W_i$ and $\iota(Z) = 0$. Clearly, F is weakly $AC_\Delta(W_i)$ for each i. Without losing generality, we can assume that $W_i \subseteq W_{i+1}$ for each i. Note that for each positive integer n, we have

$$\bigcup_{i=1}^{n} X_i \setminus \bigcup_{i=1}^{n} W_i = X_n \setminus W_n.$$

Thus, we infer that for each n,

$$0 \leq \iota \left(\bigcup_{i=1}^{n} X_i \setminus \bigcup_{i=1}^{n} W_i \right) < \frac{1}{n}.$$

It follows that $\iota(Z) = \iota \left(\overline{E} \setminus \bigcup_{i=1}^{\infty} W_i \right) = 0$ and the proof is complete.

□

We will employ the ACG_Δ condition again in Chapter 4. Next we introduce a stronger form of the ACG_Δ condition.

Let $Y \subseteq \overline{E}$ be a measurable set. Suppose $D_1 = \{(I, x)\}$ is a partial division of E with $x \in Y$ and $D_2 = \{(J, y)\}$ is a refinement of D with $y \in Y$. Then for every J such that $(J, y) \in D_2$, there exists $(I, x) \in D_1$ satisfying $J \subseteq I$. Consequently, we have

$$\bigcup_{(J,y) \in D_2} J \subseteq \bigcup_{(I,x) \in D_1} I.$$

Note that since ι is countably additive, we have

$$(D_1) \sum \iota(I) - (D_2) \sum \iota(J) \tag{3.3}$$

$$= \iota \left(\bigcup_{(I,x) \in D_1} I \right) - \iota \left(\bigcup_{(J,y) \in D_2} J \right)$$

$$= \iota \left(\bigcup_{(I,x) \in D_1} I \setminus \bigcup_{(J,y) \in D_2} J \right)$$

Likewise, if F is the primitive of an H-integrable function on E, then the equalities in (3.3) hold with ι replaced by F. Throughout the remainder of this book we shall write

$$(D_1 \backslash D_2) \sum = (D_1) \sum - (D_2) \sum.$$

As we shall see, the introduction of the notion of $D_1 \backslash D_2$ is to handle the oscillation of F outside a given measurable subset Y of $\overline{E}$. This leads to the following definition.

Definition 3.5 Let $Y \subseteq \overline{E}$ be a measurable set. An elementary-set function F is said to be $AC^\Delta(Y)$ if F is $ABRS$ on Y and for every $\varepsilon > 0$, there exist a gauge G on E and $\eta > 0$ such that for any two G-fine partial divisions $D_1 = \{(I, x)\}$ and $D_2 = \{(J, y)\}$ of E with associated points $x, y \in Y$ such that D_2 is a refinement of D_1 satisfying the condition that $(D_1 \backslash D_2) \sum \iota(I) < \eta$, we have

$$\left| (D_1 \backslash D_2) \sum F(I) \right| < \varepsilon. \tag{3.4}$$

Here D_2 may be void. If $\overline{E}$ is the union of closed sets X_i, $i = 1, 2, \ldots$ such that F is $AC^\Delta(X_i)$ for each i, then F is said to be ACG^Δ on E. If the union of X_i is a proper subset W of $\overline{E}$, then F is said to be ACG^Δ on W.

Note that since D_2 may be void, an elementary-set function which is $AC^\Delta(Y)$ as in Definition 3.5 is necessarily $AC_\Delta(Y)$ as in Definition 3.4.

The above definition is a generalisation of the ACG_δ^{**} condition given in [23; 27] which is for the real line. Comparing with the latter, we note that in our definition $ABRS$ is an additional condition. However, we remark that in the case when the topology $\mathcal{T}$ of X is induced by a metric in X, and in particular when X is the real line with $\mathcal{T}$ being the topology induced by the usual metric, a closed set $Y \subseteq \overline{E}$ is totally bounded and so the $ABRS$ condition is superfluous in the above definition. Indeed, if (3.4) holds for all partial divisions D_1 and D_2 as described in Definition 3.5 with $\varepsilon = 1$, then Y is covered by, say, N open balls with radius η and so for any G-fine division D of E we will have $(D)\sum |F(I)| \leq N$. If the topology $\mathcal{T}$ is not metrizable, the *ABRS* condition may not follow as a consequence of the AC^Δ condition.

We shall later see that the AC^Δ condition is tailored to allow a proof of the controlled convergence theorem to go through.

The following theorem indicates the relevance of the above concepts to H-integrability. Throughout the remainder of this book we shall write f_Y for $f\chi_Y$, that is, the function f_Y is given by $f_Y(x) = f(x) \times \chi_Y(x)$ for each $x \in \overline{E}$.

Theorem 3.4 *Let f be a function which is H-integrable on E with primitive F and let $Y \subseteq \overline{E}$ be a closed set. If F is $AC^\Delta(Y)$, then f_Y is absolutely H-integrable on E.*

Proof. Since for each $x \in \overline{E}$, we have

$$|f_Y(x)| = \max(f_Y(x), 0) + \max(-f_Y(x), 0),$$

it suffices to prove that $f^* := \max(f_Y, 0)$ is H-integrable on E. First, let $\varepsilon > 0$ be given and let the gauge G_1 on E and the positive number η be as described in Definition 3.5. Then for every G_1-fine interval-point pair (I, x) such that $I \subseteq E$, we let

$$F^*(I) = \begin{cases} \max(F(I), 0), & x \in Y, \\ 0, & x \notin Y. \end{cases}$$

Since f is H-integrable on E, by Henstock's lemma, we may assume that for all G_1-fine partial divisions D of E, we have

$$(D)\sum |f(x)\iota(I) - F(I)| < \varepsilon.$$

For each elementary subset E_0 of E, we define

$$\Psi(E_0) = \sup_D \left((D)\sum |f(x)\iota(I) - F(I)|\right)$$

where the supremum is taken over all G_1-fine divisions D of E_0. Note that Ψ is superadditive over elementary subsets of E and $\Psi(E_0) \leq \varepsilon$ for each elementary subset E_0 of E. Now for every G_1-fine interval-point pair (I, x) where $I \subseteq E$ and $x \in Y$, we have

$$f(x)\iota(I) \leq F(I) + \Psi(I)$$
$$\leq F^*(I) + \Psi(I)$$

which yields

$$f^*(x)\iota(I) \leq F^*(I) + \Psi(I).$$

Similarly we can show that

$$F^*(I) \leq f^*(x)\iota(I) + \Psi(I).$$

Consequently, for all G_1-fine partial divisions $D = \{(I, x)\}$ of E, we have

$$\left| (D) \sum [f^*(x)\iota(I) - F^*(I)] \right| \leq (D) \sum \Psi(I)$$
$$\leq \Psi(E)$$
$$\leq \varepsilon.$$

Now, since F is $AC^\triangle(Y)$, it is $ABRS$ on Y by definition. Hence we can modify the gauge G_1 if necessary such that for each gauge G which is finer than G_1 on E, we have

$$A_G := \sup_{D_G} \left((D_G) \sum F^*(I) \right) < \infty.$$

Here the supremum is over all G-fine divisions $D_G = \{(I, x)\}$ of E. Let

$$A = \inf_G A_G$$

where the infimum is over all gauges G which are finer than G_1. We shall prove that f^* is H-integrable on E with integral value A. To this end, we first choose an open set $U \supseteq Y$ such that $\iota(U \setminus Y) < \eta$ and choose a gauge G_2, finer than G_1 on E, such that

$$A \leq A_{G_2} < A + \varepsilon.$$

We now choose a gauge G^* which is finer than G_2 on E such that $G^*(x) \subseteq U$ when $x \in Y$ and $G^*(x) \cap Y = \emptyset$ when $x \notin Y$. Note that

$$A \leq A_{G^*} \leq A_{G_2} < A + \varepsilon.$$

We further choose a G^*-fine division $D_1 = \{(J, y)\}$ of E such that

$$A_{G^*} - \varepsilon < (D_1) \sum F^*(J) \leq A_{G^*}.$$

Modify G^* if necessary so that every G^*-fine division D of E can be partitioned into a refinement D' of D_1. For convenience we still write D for D'. Let D_1^* be the subset of D_1 such that $y \in Y$ and let $D = \{(I, x)\}$ be any G^*-fine division of E. Note that if D^* is the subset of D such that $y \in Y$ we will have

$$(D_1) \sum F^*(J) \leq (D) \sum F^*(I) + \left| (D_1^* \setminus D^*) \sum F(I) \right|.$$

Note that

$$(D_1^* \setminus D^*) \sum \iota(I) = (D_1) \sum_{y \in Y} \iota(J) - (D) \sum_{x \in Y} \iota(I)$$
$$\leq \iota(U \setminus Y)$$
$$< \eta.$$

It follows from the $AC^\Delta(Y)$ condition that

$$\left| (D_1^* \setminus D^*) \sum F(I) \right| < \varepsilon.$$

Therefore, we have

$$0 \leq A_{G^*} - (D) \sum F^*(I)$$
$$\leq A_{G^*} - (D_1) \sum F^*(J) + \left| (D_1^* \setminus D^*) \sum F(I) \right|$$
$$< 2\varepsilon.$$

Consequently, we obtain

$$\left| (D) \sum f^*(x)\iota(I) - A \right| \leq \left| (D) \sum [f^*(x)\iota(I) - F^*(I)] \right|$$
$$+ A_{G^*} - (D) \sum F^*(I)$$
$$+ A_{G^*} - A$$
$$< 4\varepsilon$$

and the proof is complete. $\qquad\qquad\qquad\qquad\qquad\square$

We next introduce the concept of equiintegrability. The idea behind this concept is that there exists a single gauge G that works for all of the functions which are H-integrable.

Definition 3.6 Let $\{f_n\}$ be a sequence of functions which are H-integrable on E. The sequence $\{f_n\}$ is H-equiintegrable on E if for every $\varepsilon > 0$ there exists a gauge G on E such that for any G-fine division $D = \{(I, x)\}$ of E and for all n, we have

$$\left| (D) \sum f_n(x)\iota(I) - (H) \int_E f_n \right| < \varepsilon.$$

Where there is no ambiguity, we will simply say $\{f_n\}$ is equiintegrable, instead of H-equiintegrable, on E.

Note that this concept does not allow one to ignore sets of measure zero. Let X be the real line with topology $\mathcal{T}$ induced by the usual metric and let $E = [0, 1]$. Consider the sequence $\{f_n\}$ of functions defined on $[0, 1]$ by $f_n(x) = 0$ for $x \in (0, 1]$ and $f_n(0) = n$. The sequence $\{f_n\}$ is equiintegrable on $(0, 1]$ but there is no gauge G on $[0, 1]$ for which

$$\left| (D) \sum f_n(x)\iota(I) \right| < 1$$

for all n whenever D is a G-fine division of $[0, 1]$.

The result we shall next prove is an important tool for H-equiintegrability. It can be proved using a slight variation of the proof for Henstock's lemma (Proposition 1.8). Here we provide a different proof.

Proposition 3.2 (Uniform Henstock's Lemma) *Suppose that the sequence $\{f_n\}$ is H-equiintegrable on E. Then for every $\varepsilon > 0$ there exists a gauge G on E, independent of n, such that for any G-fine partial division $D = \{(I, x)\}$ of E and for all n, we have*

$$(D) \sum \left| f_n(x)\iota(I) - (H) \int_I f_n \right| < \varepsilon.$$

Proof.   Given $\varepsilon > 0$, let G be a gauge on E such that for any G-fine division $D^* = \{(J, y)\}$ of E and for all n, we have

$$\left| (D^*) \sum f_n(y)\iota(J) - (H) \int_E f_n \right| < \frac{\varepsilon}{4}. \tag{3.5}$$

Let $D = \{(I, x)\}$ be any G-fine partial division of E. We shall prove that

$$\left| (D) \sum \left[f_n(x)\iota(I) - (H) \int_I f_n \right] \right| < \frac{\varepsilon}{2}. \tag{3.6}$$

First, let the component intervals of the elementary set $E \setminus \bigcup_{(I,x)\in D} I$ be $J_1, J_2, \ldots, J_m$. Since each f_n is H-integrable on E, it is H-integrable on J_j for $j = 1, 2, \ldots, m$. Then for each n and for $j = 1, 2, \ldots, m$, there is a gauge $G_n^{(j)}$, which is finer than G on J_j, and then a $G_n^{(j)}$-fine division $D_n^{(j)} = \{(J, \xi)\}$ of J_j such that

$$\left| (D_n^{(j)}) \sum f_n(\xi)\iota(J) - (H) \int_{J_j} f_n \right| < \frac{\varepsilon}{4m}.$$

Now let

$$D_0 = D \cup \left(\bigcup_{j=1}^m D_n^{(j)} \right).$$

Obviously, D_0 is a G-fine division of E and thus inequality (3.5) holds with D^* replaced by D_0. Consequently, for each $n = 1, 2, \ldots$, we have

$$\left| (D) \sum \left[f_n(x)\iota(I) - (H) \int_I f_n \right] \right|$$

$$\leq \left| (D_0) \sum f_n(x)\iota(I) - (H) \int_E f_n \right|$$

$$+ \sum_{j=1}^m \left| (D_n^{(j)}) \sum f_n(\xi)\iota(J) - (H) \int_{J_j} f_n \right|$$

$$< \frac{\varepsilon}{4} + \sum_{j=1}^m \frac{\varepsilon}{4m}$$

$$= \frac{\varepsilon}{2}.$$

Note that we have proved that inequality (3.6) holds for all G-fine partial divisions D of E. Finally, given any G-fine partial division $D = \{(I, x)\}$ of E, for $n = 1, 2, \ldots$, we let

$$D_n^+ = \left\{ (I, x) \in D : f_n(x)\iota(I) - (H) \int_I f_n > 0 \right\}$$

and $D_n^- = D \setminus D_n^+$. Applying inequality (3.6) on D_n^+ and D_n^-, we have

$$(D) \sum \left| f_n(x)\iota(I) - (H) \int_I f_n \right|$$
$$= (D_n^+) \sum \left[f_n(x)\iota(I) - (H) \int_I f_n \right]$$
$$+ \left| (D_n^-) \sum \left[f_n(x)\iota(I) - (H) \int_I f_n \right] \right|$$
$$< \frac{\varepsilon}{2} + \frac{\varepsilon}{2}$$
$$= \varepsilon$$

and this completes the proof.

$\square$

Let us introduce another concept whose relevance and relation to H-integrability is indicated in the theorem that follows the definition below.

Definition 3.7 Let $\{X_i\}$ be a sequence of closed subsets of $\overline{E}$. A sequence $\{F_n\}$ of elementary-set functions is said to satisfy the (W)-condition on $\{X_i\}$ if for each $i = 1, 2, \ldots$ and every $\varepsilon > 0$, there exist a gauge G_i on E and a positive integer N_i satisfying the condition that for any G_i-fine partial division $D = \{(I, x)\}$ of E with associated points $x \in X_i$ and whenever $n, m \geq N_i$, we have

$$\left| (D) \sum [F_n(I) - F_m(I)] \right| < \varepsilon.$$

The proof of the following theorem uses an idea in [43].

Theorem 3.5   *Let f_n, $n = 1, 2, \ldots$, be H-integrable functions on E with primitive F_n where $f_n(x) \to f(x)$ everywhere in $\overline{E}$ as $n \to \infty$ and let $\{X_i\}$ be a sequence of closed sets whose union is $\overline{E}$. If $\{F_n\}$ satisfies the (W)-condition on $\{X_i\}$ then $\{f_n\}$ is H-equiintegrable on E.*

Proof.   Let $\varepsilon > 0$ be given. Since $\{F_n\}$ satisfies the (W)-condition on $\{X_i\}$, for any positive integers i and j there exist a gauge $G_{i,j}$ on E and a positive integer $N(i, j)$ such that for any $G_{i,j}$-fine partial division $D = \{(I, x)\}$ of E with $x \in X_i$, we have

$$\left| (D) \sum [F_n(I) - F_m(I)] \right| < \frac{\varepsilon}{2^{i+j}} \tag{3.7}$$

whenever $n, m \geq N(i, j)$. Assume as we may that for each i, the sequence $\{N(i, j) : j = 1, 2, \ldots\}$ is strictly increasing. Given i, we may then choose $\{N(i+1, j) : j = 1, 2, \ldots\}$ as a subsequence of $\{N(i, j) : j = 1, 2, \ldots\}$. Consequently, $N(i+1, j) \geq N(i, j)$ for each $j = 1, 2, \ldots$ and thus the sequence $\{N(j, j) : j = 1, 2, \ldots\}$ is strictly increasing. Next, since each f_n is H-integrable on E, by Henstock's lemma, for each $n = 1, 2, \ldots$, there is a gauge G_n^* on E such that for any G_n^*-fine partial division $D = \{(I, \xi)\}$ of E we have

$$(D) \sum |f_n(\xi)\iota(I) - F_n(I)| < \frac{\varepsilon}{2^n}. \tag{3.8}$$

We may assume that G_m^* is finer than G_n^* when $m > n$. Note that any G_m^*-fine division is also G_n^*-fine though not conversely. Write $Y_1 = X_1$ and

$$Y_i = X_i \setminus \bigcup_{k=1}^{i-1} X_k$$

for $i = 2, 3, \ldots$. Then for every $\xi \in Y_i$ we choose $m(\xi) \geq N(i, i)$ such that whenever $n, k \geq m(\xi)$ we have

$$|f_n(\xi) - f_k(\xi)| < \varepsilon. \tag{3.9}$$

We may assume that $m(\xi) = N(j,j)$ for some $j > i$. Note that $N(j+1, j+1) > N(j,j)$ for each j. Now put

$$G(\xi) = G^*_{m(\xi)}(\xi) \cap G_{i,j}(\xi)$$

when $\xi \in Y_i$ for $i = 1, 2, \ldots$ and $m(\xi) = N(j,j)$. We shall show that $\{f_n\}$ is H-equiintegrable on E using the gauge G. Let $D = \{(I, \xi)\}$ be any G-fine division of E. Given n, split D into D_1 and D_2 for which $m(\xi) \geq n$ and $m(\xi) < n$ respectively. Note that when $n > m(\xi) = N(j,j)$, applying (3.7) yields

$$\left| (D_2) \sum \left[F_{m(\xi)}(I) - F_n(I) \right] \right| \tag{3.10}$$

$$= \left| \sum_{i=1}^{\infty} \sum_{j} \left(\sum_{n > m(\xi) = N(j,j), \xi \in Y_i} \left[F_{m(\xi)}(I) - F_n(I) \right] \right) \right|$$

$$< \sum_{i=1}^{\infty} \sum_{j=1}^{\infty} \frac{\varepsilon}{2^{i+j}}$$

$$= \varepsilon.$$

Consequently by applying (3.8), (3.9) and (3.10) we obtain

$$\left| (D) \sum f_n(\xi) \iota(I) - F_n(E) \right|$$

$$\leq \left| (D_1) \sum \left[f_n(\xi) \iota(I) - F_n(I) \right] \right|$$

$$+ \left| (D_2) \sum \left[f_n(\xi) \iota(I) - F_n(I) \right] \right|$$

$$< \varepsilon + \left| (D_2) \sum \left[f_n(\xi) \iota(I) - f_{m(\xi)}(\xi) \iota(I) \right] \right|$$

$$+ \left| (D_2) \sum \left[f_{m(\xi)}(\xi) \iota(I) - F_{m(\xi)}(I) \right] \right|$$

$$+ \left| (D_2) \sum \left[F_{m(\xi)}(I) - F_n(I) \right] \right|$$

$$< \varepsilon(3 + \iota(E)).$$

Hence $\{f_n\}$ is H-equiintegrable on E. □

In order to relax "everywhere" to "almost everywhere" in the above theorem, we introduce the concept of the uniform strong Lusin condition.

Definition 3.8   A sequence $\{F_n\}$ of elementary-set functions is said to satisfy the uniform strong Lusin condition, or briefly, $\{F_n\}$ is USL, if for every $\varepsilon > 0$ and every set S of measure zero, there exists a gauge G on E, independent of n, such that for any G-fine partial division $D = \{(I, \xi)\}$ of E with $\xi \in S$ and for all n, we have

$$(D) \sum |F_n(I)| < \varepsilon.$$

If $F_n = F$ for all n, we say that F satisfies the strong Lusin condition or, in brevity, F is SL.

The USL condition is employed by Lee [22] in connection with the study of the ACG^* condition, which is the real line analogue of the ACG_Δ condition. As we shall later see, the USL condition is closely related to H-equiintegrability as is seen in [29]. On the other hand, the following result shows the relevance of the strong Lusin condition to H-integrability.

Proposition 3.3   *Let f be a function which is H-integrable on E and let F be the primitive of f. Then F satisfies the strong Lusin condition.*

Proof.   Let S be a set of measure zero. For each $i = 1, 2, \ldots$, we define

$$S_i = \{x \in S : i - 1 \le |f(x)| < i\}$$

and note that S_i is measurable and $S_i \subseteq S$. Hence each S_i is a set of measure zero. Now let $\varepsilon > 0$ be given. For each i, by Condition $(*)$, we choose an open set U_i such that $S_i \subseteq U_i$ and

$$\iota(U_i) < \frac{\varepsilon}{i \times 2^{i+1}},$$

and a gauge G_i on E such that $G_i(x) \subseteq U_i$ whenever $x \in S_i$. Since f is H-integrable on E, by Henstock's lemma (Proposition 1.8), there exists a gauge G_0 on E such that for any G_0-fine partial division $D = \{(I, x)\}$ of E, we have

$$(D) \sum |f(x)\iota(I) - F(I)| < \frac{\varepsilon}{2}.$$

Define a gauge G on E such that $G(x) = G_i(x) \cap G_0(x)$ if $x \in S_i$ and $G(x) = G_0(x)$ if $x \notin S$. Let $D = \{(I, x)\}$ be a G-fine partial division of E with $x \in S$ and for each i, let
$$D_i = \{(I, x) \in D : x \in S_i\}.$$
Then we obtain
$$(D)\sum |F(I)| \leq (D)\sum |f(x)\iota(I) - F(I)| + (D)\sum |f(x)|\,\iota(I)$$
$$< \frac{\varepsilon}{2} + \sum_{i=1}^{\infty}\left[(D_i)\sum |f(x)|\,\iota(I)\right]$$
$$< \frac{\varepsilon}{2} + \sum_{i=1}^{\infty}\frac{\varepsilon}{2^{i+1}}$$
$$= \varepsilon$$
and the result follows. $\qquad\qquad\qquad\qquad\qquad\qquad\qquad\square$

With the USL condition, Theorem 3.5 can be modified as follows.

Theorem 3.6 *Let f_n, $n = 1, 2, \ldots$, be H-integrable on E with primitive F_n where $f_n(x) \to f(x)$ almost everywhere in $\overline{E}$ as $n \to \infty$ and let $\{X_i\}$ be a sequence of closed subsets of $\overline{E}$ such that $\overline{E}\backslash\bigcup_{i=1}^{\infty}X_i$ is of measure zero. If $\{F_n\}$ satisfies the (W)-condition on $\{X_i\}$ and is USL, then there is a sequence $\{f_n^*\}$ of functions such that $f_n^*(x) = f_n(x)$ almost everywhere in $\overline{E}$ for all n and which is H-equiintegrable on E.*

The proof of the above result parallels that of Theorem 3.5 except that we replace f_n with f_n^*. Furthermore, when we define a common gauge G, we put
$$G(\xi) = G_{m(\xi)}(\xi)$$
when $\xi \in Y_i$ for $i = 1, 2, \ldots$ where $\overline{E}\backslash\bigcup_{i=1}^{\infty}Y_i$ is of measure zero. On $\overline{E}\backslash\bigcup_{i=1}^{\infty}Y_i$ we define G to be that in the definition of

the USL condition. Then the result follows as in the proof of Theorem 3.5.

We have seen from the above results that in order to deal with a sequence $\{f_n\}$ of H-integrable functions, we need a condition that is uniform in n on the sequence $\{F_n\}$ of primitives of these functions. In view of this, it is essential to introduce the following definitions. We shall see in the next section that they are important conditions in proving the controlled convergence theorem.

Definition 3.9   Let $Y \subseteq \overline{E}$ be a measurable set. We say that a sequence $\{F_n\}$ of elementary-set functions is $UAC_\Delta(Y)$ (respectively weakly $UAC_\Delta(Y)$) if F_n is $AC_\Delta(Y)$ (respectively weakly $AC_\Delta(Y)$) uniformly in n, that is, the gauge G in Definition 3.4 is independent of n. If $\overline{E}$ is the union of a sequence of closed sets X_i such that $\{F_n\}$ is $UAC_\Delta(X_i)$ (respectively weakly $UAC_\Delta(X_i)$) for each i, then $\{F_n\}$ is said to be $UACG_\Delta$ (respectively weakly $UACG_\Delta$) on E. If the union of X_i is a proper subset W of $\overline{E}$, then $\{F_n\}$ is said to be $UACG_\Delta$ (respectively weakly $UACG_\Delta$) on W.

Definition 3.10   Let $Y \subseteq \overline{E}$ be a measurable set. We say that a sequence $\{F_n\}$ of elementary-set functions is $UAC^\Delta(Y)$ if F_n is $AC^\Delta(Y)$ uniformly in n, that is, the gauge G in Definition 3.5 is independent of n. If $\overline{E}$ is the union of a sequence of closed sets X_i such that $\{F_n\}$ is $UAC^\Delta(X_i)$ for each i, then $\{F_n\}$ is said to be $UACG^\Delta$ on E. If the union of X_i is a proper subset W of $\overline{E}$, then $\{F_n\}$ is said to be $UACG^\Delta$ on W.

Note that if $\{F_n\}$ is $UAC^\Delta(Y)$ (respectively $UACG^\Delta$ on W), then it is necessarily $UAC_\Delta(Y)$ (respectively $UACG_\Delta$ on W).

By virtue of Condition $(*)$, it is routine to verify that if $\{F_n\}$ is weakly $UACG_\Delta$ on E, then it satisfies the uniform strong Lusin condition on E. We still provide the proof here for completeness.

Proposition 3.4 *Let $\{F_n\}$ be a sequence of elementary-set functions on E. If $\{F_n\}$ is weakly $UACG_\Delta$ on E, then it is USL.*

Proof. Let $\{X_i\}$ be a sequence of closed sets whose union is $\overline{E}$ such that $\{F_n\}$ is weakly $UAC_\Delta(X_i)$ for each i. Without losing generality, we may assume that $X_i \subseteq X_{i+1}$ for each i. Given $\varepsilon > 0$, for each i, there exist a gauge G_i on E and $\eta_i > 0$ such that for any G_i-fine partial division $D = \{(I, x)\}$ of E with $x \in X_i$ satisfying $(D) \sum \iota(I) < \eta_i$, we have, for all n,
$$(D) \sum |F_n(I)| < \frac{\varepsilon}{2^i}.$$
By Condition $(*)$, for each i, we can choose an open subset U_i of $\overline{E}$ such that $X_i \subseteq U_i$ and $\iota(U_i) < \eta_i$. Now let $Z \subseteq \overline{E}$ be a set of measure zero. For each i, we define $Z_i = Z \cap X_i$ and note that $\bigcup_{i=1}^{\infty} Z_i = Z$. We then define a gauge G on E such that
$$G(x) \subseteq G_i(x) \cap U_i$$
if $x \in Z_i$. For any G-fine partial division $D = \{(I, x)\}$ of E such that $x \in Z$ we let
$$D_i = \{(I, x) \in D : x \in Z_i\}.$$
Clearly, if $(I, x) \in D_i$, then $I \subseteq G(x) \subseteq U_i$ and thus
$$(D_i) \sum \iota(I) \leq \iota(U_i) < \eta_i.$$
It follows that, for all n, we have
$$(D_i) \sum |F_n(I)| < \frac{\varepsilon}{2^i}.$$
Consequently, we obtain
$$(D) \sum |F_n(I)| \leq \sum_{i=1}^{\infty} \left[(D_i) \sum |F_n(I)| \right]$$
$$< \sum_{i=1}^{\infty} \frac{\varepsilon}{2^i}$$
$$= \varepsilon$$
which holds for all n, and the result follows. $\qquad\square$

It is well known that uniform convergence of a sequence $\{f_n\}$ of measurable functions on a closed set implies pointwise convergence there but not conversely. However, the well-known Egoroff's theorem (for a reference, see [17, Theorem 11.32]) gives a partial converse, namely, pointwise convergence on a closed set implies uniform convergence there except for an open set U which we can make as small as we like, but we cannot make it disappear entirely. We will need this important result in proving the last theorem of this section as well as the controlled convergence theorem in the next section. We formalise and prove this result in the following theorem.

Theorem 3.7 (Egoroff's Theorem) *Let $f_1, f_2, \ldots$ be measurable functions on $\overline{E}$. If $f_n(x) \to f(x)$ almost everywhere in $\overline{E}$ as $n \to \infty$, then for every $\eta > 0$ there is an open set U with $\iota(U) < \eta$ such that f_n converges to f uniformly on $\overline{E} \setminus U$.*

Proof. Let Y be the set of all points $x \in \overline{E}$ such that $f_n(x) \to f(x)$ as $n \to \infty$. Then $Z = \overline{E} \setminus Y$ is a set of measure zero. For each $p = 1, 2, \ldots$ and for each $q = 1, 2, \ldots$, define

$$U_{p,q} = \bigcup_{n=p}^{\infty} \left\{ x \in \overline{E} : |f_n(x) - f(x)| \geq \frac{1}{q} \right\}$$

and

$$Z_q = \bigcap_{p=1}^{\infty} U_{p,q}.$$

Clearly, $U_{p,q}$ and Z_q are measurable sets. We shall prove that $Z_q \subseteq Z$ for $q = 1, 2, \ldots$. For a fixed q, let $x \in Z_q$. Then $x \in U_{p,q}$ for every p, and so for each p there exists an integer $n \geq p$ such that

$$|f_n(x) - f(x)| \geq \frac{1}{q}.$$

Hence $f_n(x)$ does not converge to $f(x)$ as $n \to \infty$, and thus this point x belongs to Z. Thus, $Z_q \subseteq Z$ and it follows that

$$0 \leq \iota(Z_q) \leq \iota(Z) = 0$$

and so $\iota(Z_q) = 0$. With q fixed, for $p = 1, 2, \ldots$, we have

$$Z_q = \bigcap_{k=1}^{\infty} U_{k,q} \subseteq U_{p+1,q} \subseteq U_{p,q}.$$

Consequently, for $p = 1, 2, \ldots$, we obtain

$$\lim_{p \to \infty} \iota(U_{p,q}) = \iota(Z_q) = 0.$$

Now let $\eta > 0$ be given. Since $\lim_{p \to \infty} \iota(U_{p,q}) = 0$, for each positive integer q there exists a positive integer $p(q)$ such that $\iota(U_{p(q),q}) < \frac{\eta}{2^q}$. Define

$$U = \bigcup_{q=1}^{\infty} U_{p(q),q}$$

and note that

$$\iota(U) \le \sum_{q=1}^{\infty} \iota(U_{p(q),q}) < \sum_{q=1}^{\infty} \frac{\eta}{2^q} = \eta.$$

It remains to prove that f_n converges to f uniformly on $\overline{E} \setminus U$. Note that if $x \in \overline{E} \setminus U$, then $x \notin U$ and so $x \notin U_{p(q),q}$ for all q. As a result, whenever $n \ge p(q)$, we have

$$|f_n(x) - f(x)| < \frac{1}{q}.$$

Now let $\varepsilon > 0$ be given and choose a positive integer q_0 large enough so that $\frac{1}{q_0} < \varepsilon$. Then whenever $n \ge p(q_0)$, we have, for any $x \in \overline{E} \setminus U$,

$$|f_n(x) - f(x)| < \frac{1}{q_0} < \varepsilon.$$

Note that by Condition $(*)$, we can choose an open set $U^* \supseteq U$ such that $\iota(U^*) < \eta$. This completes the proof. $\qquad \square$

The next theorem we shall prove shows the relation of H-equiintegrability to ACG_Δ. We will need the following lemma and the equiintegrability theorem which we will prove in Section 3.3.

Lemma 3.1 *Let f be a real-valued function on $\overline{E}$ and let Z be a subset of $\overline{E}$ such that $\iota(Z) = 0$. Then for every $\varepsilon > 0$, there exists a gauge G on E such that for any G-fine partial division $D = \{(I, x)\}$ of E with $x \in Z$, we have*

$$(D) \sum |f(x)| \iota(I) < \varepsilon.$$

Proof. For each $i = 1, 2, \ldots$, let $Z_i = \{x \in Z : i - 1 \leq |f(x)| < i\}$. Obviously, $\iota(Z_i) = 0$ for each i. Now let $\varepsilon > 0$ be given. By Condition $(*)$, for each i, we can find an open set U_i such that $Z_i \subseteq U_i$ and

$$\iota(U_i) < \frac{\varepsilon}{i \times 2^i}.$$

Define a gauge G on E such that for each i, if $x \in Z_i$, then $G(x) \subseteq U_i$. Let $D = \{(I, x)\}$ be a G-fine partial division of E with $x \in Z$. Note that for each $(I, x) \in D$, we must have $x \in Z_i$ for some i and $I \subseteq U_i$. Thus, for each i, we have

$$(D) \sum_{x \in Z_i} \iota(I) \leq \iota(U_i) < \frac{\varepsilon}{i \times 2^i}.$$

It follows that

$$
\begin{aligned}
(D) \sum |f(x)| \iota(I) &= \sum_{i=1}^{\infty} \left[(D) \sum_{x \in Z_i} |f(x)| \iota(I) \right] \\
&< \sum_{i=1}^{\infty} \left[(D) \sum_{x \in Z_i} \iota(I) \right] i \\
&< \sum_{i=1}^{\infty} \left(\frac{\varepsilon}{i \times 2^i} \right) i \\
&= \varepsilon.
\end{aligned}
$$

The proof is complete. $\square$

Theorem 3.8 *Let $f_1, f_2, \ldots$ be H-integrable functions on E and let F_n be the primitive of f_n for each n. Suppose that $\{f_n\}$ is H-equiintegrable on E and $f_n(x) \to f(x)$ everywhere in $\overline{E}$ as $n \to \infty$. Then $\{F_n\}$ is weakly $UACG_\Delta$ on $E \setminus Z$ where $\iota(Z) = 0$, and is USL.*

Proof. Since $f_n(x) \to f(x)$ everywhere in $\overline{E}$ as $n \to \infty$ and $\{f_n\}$ is H-equiintegrable on E, by the equiintegrability theorem (Theorem 3.10), the function f is H-integrable on E. Let F denote the primitive of f. By applying Henstock's lemma on f and uniform Henstock's lemma (Proposition 3.2) on $\{f_n\}$, there exists a gauge G_0 on E such that for any G_0-fine partial division $D = \{(I, x)\}$ of E, we have

$$(D) \sum |f(x)\iota(I) - F(I)| < \frac{\varepsilon}{4} \qquad (3.11)$$

and

$$(D) \sum |f_n(x)\iota(I) - F_n(I)| < \frac{\varepsilon}{4}. \qquad (3.12)$$

Since f is H-integrable on E, by Proposition 3.1, its primitive F is weakly ACG_Δ on $\overline{E} \setminus Z_0$ where $\iota(Z_0) = 0$. Thus, for each i, there exist a measurable subset X_i of $\overline{E}$, a gauge G_i on E, and $\eta_i > 0$ such that

$$\overline{E} \setminus Z_0 = \bigcup_{i=1}^{\infty} X_i,$$

where $X_i \cap X_j = \emptyset$ for $i \neq j$, and that for each i, whenever $D = \{(I, x)\}$ is a G_i-fine partial division of E with $x \in X_i$ satisfying $(D) \sum \iota(I) < \eta_i$, we have

$$(D) \sum |F(I)| < \frac{\varepsilon}{4}. \qquad (3.13)$$

Next, since each f_n is H-integrable on E, it is measurable. Then since $f_n \to f$ as $n \to \infty$, by Egoroff's theorem (Theorem 3.7), for each positive integer k, there exsts a measurable set $W_k \subseteq \overline{E}$

such that $f_n \to f$ uniformly on W_k as $n \to \infty$ and $\iota(\overline{E}\backslash W_k) < \frac{1}{k}$. Since for each k,

$$0 \leq \iota\left(\overline{E}\backslash \bigcup_{j=1}^{\infty} W_j\right) \leq \iota(\overline{E}\backslash W_k),$$

it follows that $\iota\left(\overline{E}\backslash \bigcup_{k=1}^{\infty} W_k\right) = 0$. Let

$$Z_1 = Z_0 \cup \left(\overline{E}\backslash \bigcup_{k=1}^{\infty} W_k\right)$$

and note that $\iota(Z_1) = 0$. For $i = 1, 2, \ldots$ and $k = 1, 2, \ldots$, let $N_{i,k}$ be a positive integer such that

$$|f_n(x) - f(x)| < \frac{\varepsilon}{4\eta_i} \tag{3.14}$$

whenever $x \in X_i \cap W_k$ and $n \geq N_{i,k}$. Define

$$Y_{i,0} = X_i \backslash \bigcup_{k=1}^{\infty} W_k \quad \text{for} \quad i = 1, 2, \ldots;$$

and

$$Y_{i,k} = X_i \cap W_k \quad \text{for} \quad i = 1, 2, \ldots \text{ and } k = 1, 2, \ldots.$$

Clearly,

$$\bigcup_{i=1}^{\infty} \bigcup_{k=0}^{\infty} Y_{i,k} = \bigcup_{i=1}^{\infty} X_i = \overline{E}\backslash Z_0.$$

Moreover, since for each i we have

$$0 \leq \iota(Y_{i,0}) \leq \iota\left(\overline{E}\backslash \bigcup_{k=1}^{\infty} W_k\right) = 0,$$

we infer that $\iota(Y_{i,0}) = 0$ for each i. Define a function f^* on $\overline{E}$ given by

$$f^*(x) = \sup_n |f_n(x)|$$

for each $x \in \overline{E}$, where the supremum is over all positive integers n. Apply Lemma 3.1 on f^* and each $Y_{i,0}$ to obtain for $i = 1, 2, \ldots$, a gauge $G^{(i)}$ on E such that whenever $D = \{(I, x)\}$ is a $G^{(i)}$-fine partial division of E with $x \in Y_{i,0}$, we have

$$(D) \sum |f^*(x)| \iota(I) < \frac{3\varepsilon}{4}. \tag{3.15}$$

We shall prove that $\{F_n\}$ is weakly UAC_Δ on each $Y_{i,k}$. To this end, select a gauge G on E so that if $x \in Y_{i,0}$ for $i = 1, 2, \ldots$, then

$$G(x) \subseteq G_0(x) \cap G^{(i)}(x),$$

and if $x \in Y_{i,k}$ for $i = 1, 2, \ldots$ and $k = 1, 2, \ldots$, then

$$G(x) \subseteq G_0(x) \cap G_i(x).$$

Now for fixed positive integers i and k, we let $n \geq N_{i,k}$. Then for any G-fine partial division $D = \{(I, x)\}$ of E with $x \in Y_{i,k}$ satisfying $(D) \sum \iota(I) < \eta_i$, we obtain, by applying inequalities (3.12), (3.14), (3.11) and (3.13) in that order,

$$\begin{aligned}
(D) \sum |F_n(I)| &\leq (D) \sum |f_n(x)\iota(I) - F_n(I)| \\
&\quad + (D) \sum |f_n(x)\iota(I) - f(x)\iota(I)| \\
&\quad + (D) \sum |f(x)\iota(I) - F(I)| \\
&\quad + (D) \sum |F(I)| \\
&< \frac{\varepsilon}{4} + \left(\frac{\varepsilon}{4\eta_i}\right)\eta_i + \frac{\varepsilon}{4} + \frac{\varepsilon}{4} \\
&= \varepsilon.
\end{aligned}$$

Hence $\{F_n\}$ is weakly UAC_Δ on each $Y_{i,k}$, for $i = 1, 2, \ldots$ and $k = 1, 2, \ldots$. It remains to show that $\{F_n\}$ is weakly UAC_Δ on each $Y_{i,0}$. For a fixed i and for all n, whenever $D = \{(I, x)\}$ is a $G^{(i)}$-fine partial division of E with $x \in Y_{i,0}$, inequalities (3.12) and (3.15) yield

$$(D) \sum |F_n(I)| \leq (D) \sum |f_n(x)\iota(I) - F_n(I)|$$
$$+(D) \sum |f_n(x)|\,\iota(I)$$
$$< \frac{\varepsilon}{4} + \frac{3\varepsilon}{4}$$
$$= \varepsilon$$

as desired. Finally, for each $i = 1, 2, \ldots$, and each $k = 0, 1, \ldots$, we choose a closed set $V_{i,k}$ such that $V_{i,k} \subseteq Y_{i,k}$ and $\iota(Y_{i,k} \setminus V_{i,k}) < \dfrac{1}{2^{i+k}}$. Clearly,

$$\iota\left(\bigcup_{i=1}^{\infty}\bigcup_{k=0}^{\infty} V_{i,k}\right) = \iota\left(\bigcup_{i=1}^{\infty}\bigcup_{k=0}^{\infty} Y_{i,k}\right).$$

Let

$$Z = \overline{E} \setminus \bigcup_{i=1}^{\infty}\bigcup_{k=0}^{\infty} V_{i,k}$$

and note that

$$\iota(Z) = \iota\left(\overline{E} \setminus \bigcup_{i=1}^{\infty}\bigcup_{k=0}^{\infty} Y_{i,k}\right)$$
$$= \iota\left(\overline{E} \setminus (\overline{E} \setminus Z_0)\right)$$
$$= \iota(Z_0)$$
$$= 0.$$

Therefore $\{F_n\}$ is weakly $UACG_\Delta$ on $\overline{E} \setminus Z$. Similar to how we have just proved that $\{F_n\}$ is weakly UAC_Δ on each $Y_{i,0}$, to prove that $\{F_n\}$ is USL, we let a set S of measure zero be given and apply Lemma 3.1 on f^* and S to obtain a gauge G_S on E such that whenever $D = \{(I, x)\}$ is a G_S-fine partial division of E with $x \in S$, we have

$$(D) \sum |f_n(x)|\,\iota(I) < \frac{3\varepsilon}{4} \qquad (3.16)$$

for all n. Consequently, by inequalities (3.12) and (3.16) we obtain

$$(D)\sum |F_n(I)| < \varepsilon$$

and the proof is complete.

□

As in the case of Theorem 3.6, in the above theorem we can relax the requirement on the convergence of f_n to f from "everywhere" to "almost everywhere" by imposing the USL condition and obtain the following result.

Corollary 3.2 *Let $f_1, f_2, \ldots$ be H-integrable functions on E and let F_n be the primitive of f_n for each n. Suppose that $\{f_n\}$ is H-equiintegrable on E and $f_n(x) \to f(x)$ almost everywhere in $\overline{E}$ as $n \to \infty$. If $\{F_n\}$ is USL, then it is weakly $UACG_\Delta$ on $\overline{E} \setminus Z$ where $\iota(Z) = 0$.*

We shall now prove a partial converse of the preceding result. Our proof here generalises that of a similar result in [2] which is for the real line.

Theorem 3.9 *Let $f_1, f_2, \ldots$ be H-integrable functions on E and let F_n be the primitive of f_n for each n. Suppose that $f_n(x) \to f(x)$ almost everywhere in $\overline{E}$ as $n \to \infty$ and $\{F_n\}$ is $UACG^\Delta$ on $\overline{E} \setminus Z$, where $\iota(Z) = 0$, and is USL, then $\{f_n\}$ is H-equiintegrable on E.*

Proof. Let $f_n(x) \to f(x)$ as $n \to \infty$ for every x in $\overline{E} \setminus Z_0$ where $\iota(Z_0) = 0$ and let $\{X_i\}$ be a monotone increasing sequence of closed sets whose union is $\overline{E} \setminus Z$ such that $\{F_n\}$ is $UAC^\Delta(X_i)$ for each i. As is seen in the proof of Theorem 3.8, by Egoroff's theorem (Theorem 3.7), there exist mutually disjoint closed sets Y_j, $j = 1, 2, \ldots$, such that

$$\iota\left(\overline{E} \setminus \bigcup_{j=1}^{\infty} Y_j\right) = 0$$

and the sequence $\{f_n\}$ is uniformly convergent on each Y_j. Set

$$Z^* = Z \cup Z_0 \cup \left(\overline{E} \setminus \bigcup_{j=1}^{\infty} Y_j \right)$$

and $W_{i,j} = X_i \cap Y_j$ for each $i = 1, 2, \ldots$ and $j = 1, 2, \ldots$. Note that Z^* is a set of measure zero. Now given $\varepsilon > 0$ and positive integers i and j, there exists a positive integer $N_{i,j}$ such that

$$|f_m(x) - f_n(x)| < \frac{\varepsilon}{21 \times \iota(E) \times 2^{i+j}} \tag{3.17}$$

for each $x \in W_{i,j}$ and $m, n \geq N_{i,j}$. Since $\{F_n\}$ is $UAC^{\Delta}(X_i)$ for each i and thus is $UAC^{\Delta}(W_{i,j})$ for each i and j, there exist a gauge $G^{(1)}$ on E, which is independent of n, and $\eta_{i,j} > 0$ such that whenever $D = \{(I, x)\}$ and $D^* = \{(I^*, x^*)\}$ are $G^{(1)}$-fine partial divisions of E with associated points $x, x^* \in W_{i,j}$, where D^* is a refinement of D satisfying $(D \setminus D^*) \sum \iota(I) < \eta_{i,j}$, we have

$$\left| (D \setminus D^*) \sum F_n(I) \right| < \frac{\varepsilon}{21 \times 2^{i+j}} \tag{3.18}$$

for all n. Define a function f^* on $\overline{E}$ given by

$$f^*(x) = \sup_n |f_n(x)|$$

if $x \in Z^*$ and 0 otherwise. Since $\{F_n\}$ is USL and $\iota(Z^*) = 0$, there exists a gauge $G^{(2)}$ on E, which is independent of n, such that whenever $D = \{(I, x)\}$ is a $G^{(2)}$-fine partial division of E with associated points $x \in Z^*$, we have

$$(D) \sum |F_n(I)| < \frac{\varepsilon}{6}. \tag{3.19}$$

By applying Lemma 3.1 on f^*, we can assume that for the same $G^{(2)}$-fine partial division D of E with $x \in Z^*$, we also have

$$(D) \sum |f^*(x)| \iota(I) < \frac{\varepsilon}{6}. \tag{3.20}$$

Next, since each f_n is H-integrable on E, by Henstock's lemma, for positive integers n, i, and j, there exists a gauge G_n on E

such that for any G_n-fine partial division $D = \{(I, x)\}$ of E with $x \in W_{i,j}$, we have

$$(D) \sum |f_n(x)\iota(I) - F_n(I)| < \frac{\varepsilon}{21 \times 2^{i+j}}. \qquad (3.21)$$

Without losing generality, we may assume that G_{n+1} is finer than G_n for each n. For positive integers i and j, we choose an open set $U_{i,j}$ such that $W_{i,j} \subseteq U_{i,j}$ and $\iota(U_{i,j} \setminus W_{i,j}) < \eta_{i,j}$. We then define a gauge G on E such that if $x \in W_{i,j}$, we have

$$G(x) \subseteq G^{(1)}(x) \cap G_{N_{i,j}}(x) \cap U_{i,j}$$

and if $x \in Z^*$, we have $G(x) \subseteq G^{(2)}(x)$. Since each $W_{i,j}$ is closed, we can modify G if necessary such that if $x \in U_{i,j} \setminus W_{i,j}$, then $G(x) \subseteq U_{i,j} \setminus W_{i,j}$. As a result, if (I, x) is G-fine and $x \in U_{i,j} \setminus W_{i,j}$, we have

$$I \cap W_{i,j} = \emptyset.$$

Let $D = \{(I_k, x_k) : k = 1, 2, \ldots, m\}$ be a G-fine division of E and n be a fixed positive integer. We first note that

$$\left| \sum_{i=1}^{m} f_n(x_k)\iota(I_k) - F_n(E) \right|$$

$$= \left| \sum_{i=1}^{m} [f_n(x_k)\iota(I_k) - F_n(I_k)] \right|$$

$$\leq \left| \sum_{x_k \in Z^*} [f_n(x_k)\iota(I_k) - F_n(I_k)] \right|$$

$$+ \left| \sum_{i=1}^{\infty} \sum_{N_{i,j} \geq n} \sum_{x_k \in W_{i,j}} [f_n(x_k)\iota(I_k) - F_n(I_k)] \right|$$

$$+ \left| \sum_{i=1}^{\infty} \sum_{N_{i,j} < n} \sum_{x_k \in W_{i,j}} [f_n(x_k)\iota(I_k) - F_n(I_k)] \right|$$

where $\displaystyle\sum_{N_{i,j}\geq n}$ sums over all indices j such that $N_{i,j} \geq n$ and $\displaystyle\sum_{N_{i,j}<n}$ sums over all indices j such that $N_{i,j} < n$. By inequalities (3.19) and (3.20), we have

$$\left|\sum_{x_k\in Z^*} [f_n(x_k)\iota(I_k) - F_n(I_k)]\right|$$

$$\leq \sum_{x_k\in Z^*} |f_n(x_k)\iota(I_k) - F_n(I_k)|$$

$$\leq \sum_{x_k\in Z^*} |f^*(x_k)|\,\iota(I_k) + \sum_{x_k\in Z^*} |F_n(I_k)|$$

$$< \frac{\varepsilon}{3}.$$

On the other hand, note that if $x_k \in W_{i,j}$ where $N_{i,j} \geq n$ then an interval-point pair (I, x) which is G-fine is $G_{N_{i,j}}$-fine and so is G_n-fine. Thus, by inequality (3.21), we obtain

$$\left|\sum_{i=1}^{\infty} \sum_{N_{i,j}\geq n} \sum_{x_k\in W_{i,j}} [f_n(x_k)\iota(I_k) - F_n(I_k)]\right|$$

$$\leq \sum_{i=1}^{\infty} \sum_{N_{i,j}\geq n} \sum_{x_k\in W_{i,j}} |f_n(x_k)\iota(I_k) - F_n(I_k)|$$

$$< \sum_{i=1}^{\infty}\sum_{j=1}^{\infty} \frac{\varepsilon}{21 \times 2^{i+j}}$$

$$= \frac{\varepsilon}{21}.$$

It remains to prove that

$$\left|\sum_{i=1}^{\infty} \sum_{N_{i,j}<n} \sum_{x_k\in W_{i,j}} [f_n(x_k)\iota(I_k) - F_n(I_k)]\right| < \frac{\varepsilon}{3}.$$

Let G^* be a gauge on E which is finer than both G and G_n. For each $1 \leq k \leq m$ such that $x_k \in W_{i,j}$, we apply Cousin's lemma to find a G^*-fine division $D_{(k)} = \{(J, y)\}$ of I_k. We then define the partial divisions

$$D_{(k)}^+ = \{(J, y) \in D_{(k)} : y \in W_{i,j}\}$$

and

$$D_{(k)}^- = \{(J, y) \in D_{(k)} : y \notin W_{i,j}\}.$$

We further define

$$D_1 = \{(I_k, x_k) : x_k \in W_{i,j}, 1 \le k \le m\}$$

and

$$D_2 = \bigcup_{x_k \in W_{i,j}} D_{(k)}^+.$$

Clearly D_1 and D_2 are $G^{(1)}$-fine partial divisions of E with associated points in $W_{i,j}$ where D_2 is a refinement of D_1. Next, let

$$E_1 = \bigcup_{(I,x) \in D_1} I \quad \text{and} \quad E_2 = \bigcup_{(J,y) \in D_2} J$$

and note that

$$E_1 = \bigcup_{x_k \in W_{i,j}} \bigcup_{(J,y) \in D_{(k)}} J$$

and

$$E_2 = \bigcup_{x_k \in W_{i,j}} \bigcup_{(J,y) \in D_{(k)}^+} J.$$

Obviously, $E_2 \subseteq E_1$ and

$$E_1 \setminus E_2 = \bigcup_{x_k \in W_{i,j}} \bigcup_{(J,y) \in D_{(k)}^-} J.$$

Now note that whenever $x_k \in W_{i,j}$ we have

$$I_k \subseteq G(x_k) \subseteq U_{i,j} \setminus W_{i,j}$$

and whenever $(J, y) \in D_{(k)}^-$ we have

$$J \subseteq G^*(y) \subseteq G(y) \subseteq U_{i,j} \setminus W_{i,j}.$$

Hence $E_1 \setminus E_2 \subseteq U_{i,j} \setminus W_{i,j}$ and thus

$$
\begin{aligned}
(D_1 \setminus D_2) \sum \iota(J) &= (D_1) \sum \iota(I) - (D_2) \sum \iota(J) \\
&= \iota(E_1) - \iota(E_2) \\
&= \iota(E_1 \setminus E_2) \\
&\leq \iota(U_{i,j} \setminus W_{i,j}) \\
&< \eta_{i,j}.
\end{aligned}
$$

By virtue of (3.18), we obtain

$$
\left| (D_1 \setminus D_2) \sum F_n(I) \right| < \frac{\varepsilon}{21 \times 2^{i+j}}
$$

which holds for all n, where

$$
\begin{aligned}
(D_1 \setminus D_2) \sum F_n(I) &= (D_1) \sum F_n(I) - (D_2) \sum F_n(J) \\
&= F_n(E_1) - F_n(E_2) \\
&= F_n(E_1 \setminus E_2) \\
&= \sum_{x_k \in W_{i,j}} \sum_{(J,y) \in D^-_{(k)}} F_n(J).
\end{aligned}
$$

Therefore, for all n we have

$$
\left| \sum_{x_k \in W_{i,j}} \sum_{(J,y) \in D^-_{(k)}} F_n(J) \right| < \frac{\varepsilon}{21 \times 2^{i+j}}
$$

which implies that

$$
\left| \sum_{N_{i,j} < n} \sum_{x_k \in W_{i,j}} \sum_{(J,y) \in D^-_{(k)}} F_n(J) \right| < \sum_{j=1}^{\infty} \frac{\varepsilon}{21 \times 2^{i+j}} \qquad (3.22)
$$

$$
= \frac{\varepsilon}{21 \times 2^i}
$$

holds for all n. In particular, we get

$$\left| \sum_{N_{i,j} < n} \sum_{x_k \in W_{i,j}} \sum_{(J,y) \in D_{(k)}^-} F_{N_{i,j}}(J) \right| < \frac{\varepsilon}{21 \times 2^i}. \tag{3.23}$$

Also note that by inequality (3.17), we have

$$\sum_{N_{i,j} < n} \sum_{x_k \in W_{i,j}} \left| f_n(x_k) - f_{N_{i,j}}(x_k) \right| \iota(I_k) < \sum_{j=1}^{\infty} \frac{\varepsilon}{21 \times 2^{i+j}}$$

$$= \frac{\varepsilon}{21 \times 2^i}$$

and by the same token, we obtain

$$\sum_{N_{i,j} < n} \sum_{x_k \in W_{i,j}} \sum_{(J,y) \in D_{(k)}^+} \left| f_n(y) - f_{N_{i,j}}(y) \right| \iota(J) < \frac{\varepsilon}{21 \times 2^i}.$$

Furthermore, by inequality (3.21), we have

$$\sum_{N_{i,j} < n} \sum_{x_k \in W_{i,j}} \left| f_{N_{i,j}}(x_k) \iota(I_k) - F_{N_{i,j}}(I_k) \right| < \sum_{j=1}^{\infty} \frac{\varepsilon}{21 \times 2^{i+j}}$$

$$= \frac{\varepsilon}{21 \times 2^i}$$

and in a similar manner, we obtain

$$\sum_{N_{i,j} < n} \sum_{x_k \in W_{i,j}} \sum_{(J,y) \in D_{(k)}^+} \left| f_{N_{i,j}}(y) \iota(J) - F_{N_{i,j}}(J) \right| < \frac{\varepsilon}{21 \times 2^i},$$

and

$$\sum_{N_{i,j} < n} \sum_{x_k \in W_{i,j}} \sum_{(J,y) \in D_{(k)}} \left| f_{N_{i,j}}(x_k) \iota(I_k) - F_{N_{i,j}}(I_k) \right| < \frac{\varepsilon}{21 \times 2^i}.$$

Then using the above five inequalities in conjunction with inequalities (3.22) and (3.23), we obtain

$$\left| \sum_{i=1}^{\infty} \sum_{N_{i,j}<n} \sum_{x_k \in W_{i,j}} [f_n(x_k)\iota(I_k) - F_n(I_k)] \right|$$

$$\leq \sum_{i=1}^{\infty} \sum_{N_{i,j}<n} \sum_{x_k \in W_{i,j}} \left| f_n(x_k) - f_{N_{i,j}}(x_k) \right| \iota(I_k)$$

$$+ \sum_{i=1}^{\infty} \sum_{N_{i,j}<n} \sum_{x_k \in W_{i,j}} \left| f_{N_{i,j}}(x_k)\iota(I_k) - F_{N_{i,j}}(I_k) \right|$$

$$+ \sum_{i=1}^{\infty} \sum_{N_{i,j}<n} \sum_{x_k \in W_{i,j}} \sum_{(J,y) \in D_{(k)}^{+}} \left| f_{N_{i,j}}(y)\iota(J) - F_{N_{i,j}}(J) \right|$$

$$+ \sum_{i=1}^{\infty} \left| \sum_{N_{i,j}<n} \sum_{x_k \in W_{i,j}} \sum_{(J,y) \in D_{(k)}^{-}} F_{N_{i,j}}(J) \right|$$

$$+ \sum_{i=1}^{\infty} \sum_{N_{i,j}<n} \sum_{x_k \in W_{i,j}} \sum_{(J,y) \in D_{(k)}^{+}} \left| f_n(y) - f_{N_{i,j}}(y) \right| \iota(J)$$

$$+ \sum_{i=1}^{\infty} \sum_{N_{i,j}<n} \sum_{x_k \in W_{i,j}} \sum_{(J,y) \in D_{(k)}^{+}} \left| f_{N_{i,j}}(y)\iota(J) - F_{N_{i,j}}(J) \right|$$

$$+ \sum_{i=1}^{\infty} \left| \sum_{N_{i,j}<n} \sum_{x_k \in W_{i,j}} \sum_{(J,y) \in D_{(k)}^{-}} F_n(J) \right|$$

$$< 7 \times \sum_{i=1}^{\infty} \frac{\varepsilon}{21 \times 2^i}$$

$$= \frac{\varepsilon}{3}.$$

We have therefore proved that there exists a gauge G on E such that for any G-fine division $D = \{(I_k, x_k) : k = 1, 2, \ldots, m\}$ of

E, we have

$$\left| \sum_{i=1}^{m} f_n(x_k)\iota(I_k) - F_n(E) \right| < \varepsilon$$

which holds for all n. Hence $\{f_n\}$ is H-equiintegrable on E.

$\square$

3.3   The Controlled Convergence Theorem

The controlled convergence theorem [21, page 36] is arguably the best convergence theorem possible for the Henstock–Kurzweil integral in the sense that if we begin with some elementary functions, say, continuous functions, and apply controlled convergence, then we generate all the Henstock–Kurzweil integrable functions. In fact, it has also been proved for other integrals, such as the approximately continuous Perron integral [18]. This result has been generalised to the division space by Ding and Lee in [7]. In this section we shall prove among other convergence theorems the controlled convergence theorem for the H-integral.

We shall first prove the equiintegrability theorem and the basic convergence theorem, and then from each of which we deduce the generalised mean convergence theorem independently. With the aid of the basic convergence theorem, we shall prove the controlled convergence theorem.

The equiintegrability theorem was first proved by Kurzweil [20, page 40]; see also [11]. Jarnik and Kurzweil [18], and Wang [43] later proved independently that the equiintegrability theorem is equivalent to the controlled convergence theorem in the sense that if a sequence of functions is convergent in one sense then there is a subsequence which is convergent in the other sense, and vice versa.

We shall begin with the proof of the following equiintegrability theorem for the H-integral.

Theorem 3.10 (Equiintegrability Theorem) *Let the functions f_n, $n = 1, 2, \ldots$, be H-integrable on E where $f_n(x) \to f(x)$, everywhere in $\overline{E}$ as $n \to \infty$. If $\{f_n\}$ is H-equiintegrable on E, then f is H-integrable on E and*

$$\lim_{n \to \infty} (H) \int_E f_n = (H) \int_E f.$$

Proof. Let $\varepsilon > 0$ be given. Since $\{f_n\}$ is H-equiintegrable on E, there exists a division D_0 of E such that

$$\left| (D_0) \sum f_n(x)\iota(I) - (H) \int_E f_n \right| < \varepsilon$$

for all n. Choose a positive integer N so that

$$\left| (D_0) \sum f_n(x)\iota(I) - (D_0) \sum f_m(x)\iota(I) \right| < \varepsilon$$

for all $m, n \geq N$ and note that

$$\left| (H) \int_E f_n - (H) \int_E f_m \right|$$
$$\leq \left| (H) \int_E f_n - (D_0) \sum f_n(x)\iota(I) \right|$$
$$+ \left| (D_0) \sum f_n(x)\iota(I) - (D_0) \sum f_m(x)\iota(I) \right|$$
$$+ \left| (D_0) \sum f_m(x)\iota(I) - (H) \int_E f_m \right|$$
$$< 3\varepsilon.$$

Hence $\left\{ (H) \int_E f_n \right\}_{n=1}^{\infty}$ is a Cauchy sequence of real numbers. Let A be the limit of this sequence. Choose a gauge G on E such that for all G-fine divisions D of E, we have, for all n,

$$\left| (D) \sum f_n(x)\iota(I) - (H) \int_E f_n \right| < \varepsilon$$

and choose a positive integer N_1 so that

$$\left|(H)\int_E f_n - A\right| < \varepsilon$$

for all $n \geq N_1$. Now let $D = \{(I, x)\}$ be a G-fine division of E. Then there exists $k \geq N_1$ such that

$$\left|(D)\sum f(x)\iota(I) - (D)\sum f_k(x)\iota(I)\right| < \varepsilon.$$

Hence

$$\left|(D)\sum f(x)\iota(I) - A\right| \leq \left|(D)\sum f(x)\iota(I) - (D)\sum f_k(x)\iota(I)\right|$$
$$+ \left|(D)\sum f_k(x)\iota(I) - (H)\int_E f_k\right|$$
$$+ \left|(H)\int_E f_k - A\right|$$
$$< 3\varepsilon.$$

It follows that f is H-integrable on E and

$$(H)\int_E f = A = \lim_{n\to\infty}(H)\int_E f_n$$

as desired.

$\square$

We remark that "everywhere" in the above theorem cannot be replaced by "almost everywhere" as shown by an example in [11]. However, as in the case of Theorem 3.5, we can relax "everywhere" to "almost everywhere" by imposing the USL condition. The above theorem can thus be modified as follows.

Theorem 3.11 *Let the functions f_n, $n = 1, 2, \ldots$, be H-integrable on E with primitive F_n where $f_n(x) \to f(x)$ almost everywhere in $\overline{E}$ as $n \to \infty$. If $\{f_n\}$ is H-equiintegrable on E and $\{F_n\}$ is USL, then f is H-integrable on E and*

$$\lim_{n\to\infty}(H)\int_E f_n = (H)\int_E f.$$

Proof. We can choose f_n^* and f^* such that $f_n^*(x) = f_n(x)$ almost everywhere in $\overline{E}$ for all n while $f^*(x) = f(x)$ almost everywhere in $\overline{E}$, and

$$f_n^*(x) \to f^*(x)$$

everywhere in $\overline{E}$ as $n \to \infty$. In view of the USL condition, the sequence $\{f_n^*\}$ is also H-equiintegrable on E. Hence the result follows from Theorem 3.10.

$\square$

With Theorem 3.6 and Theorem 3.11, we have thus proved the following generalised mean convergence theorem for the H-integral.

Theorem 3.12 (Generalised Mean Convergence Theorem) *Let* $f_n, \ n = 1, 2, \ldots,$ *be* H-*integrable on* E *with primitive* F_n *and let* $\{X_i\}$ *be a sequence of closed subsets of* $\overline{E}$ *with* $\overline{E} \setminus \bigcup_{i=1}^{\infty} X_i$ *being of measure zero. Suppose that the following conditions are satisfied.*

(i) $f_n(x) \to f(x)$ *almost everywhere in* $\overline{E}$ *as* $n \to \infty$ *and* $\{F_n\}$ *is* USL;

(ii) *There is an elementary-set function* F *such that for each* $i = 1, 2, \ldots$ *and every* $\varepsilon > 0$ *there exist a gauge* G_i *on* E *and a positive integer* N_i *satisfying the condition that for any* G_i-*fine partial division* $D = \{(I, x)\}$ *of* E *with* $x \in X_i$ *we have, for all* $n \geq N_i$,

$$\left| (D) \sum [F_n(I) - F(I)] \right| < \varepsilon.$$

Then f *is* H-*integrable on* E *and*

$$\lim_{n \to \infty} (H) \int_E f_n = (H) \int_E f.$$

We observe that in the above theorem, with hypothesis (ii), the sequence $\{F_n\}$ satisfies the (W)-condition on $\{X_i\}$.

Our next task is to prove the basic convergence theorem and show that the above generalised mean convergence theorem can be derived as a consequence. With the aid of the basic convergence theorem, the controlled convergence theorem will be proved in a series of lemmas.

Theorem 3.13 (Basic Convergence Theorem). *Let f_n, $n = 1, 2, \ldots$, be H-integrable on E with primitive F_n where $f_n(x) \to f(x)$ almost everywhere in $\overline{E}$ as $n \to \infty$ and let F be an elementary-set function. Then in order that f is H-integrable on E with primitive F, it is necessary and sufficient that for every $\varepsilon > 0$ there is a function $M(x)$ taking integer values such that for infinitely many $m(x) \geq M(x)$ there is a gauge G on E satisfying the condition that for any G-fine division $D = \{(I, x)\}$ of E, we have*

$$\left| (D) \sum F_{m(x)}(I) - F(E) \right| < \varepsilon.$$

Proof. As usual, we may assume that $f_n(x) \to f(x)$ everywhere in $\overline{E}$ as $n \to \infty$. Suppose f is H-integrable on E with primitive F. Given $\varepsilon > 0$ and $x \in \overline{E}$, there is an integer $M(x)$ such that whenever $m(x) \geq M(x)$,

$$|f_{m(x)}(x) - f(x)| < \varepsilon.$$

Since each f_n is H-integrable on E, there is a gauge G_n on E such that for any G_n-fine division $D = \{(I, x)\}$ of E, we have

$$(D) \sum |F_n(I) - f_n(x)\iota(I)| < \frac{\varepsilon}{2^n}.$$

Also, there is a gauge G_0 on E such that for any G_0-fine division $D = \{(I, x)\}$ of E, we have

$$(D) \sum |F(I) - f(x)\iota(I)| < \varepsilon.$$

Now for every $m(x) \geq M(x)$, we can find a gauge G on E such that

$$G(x) \subseteq G_{m(x)}(x) \cap G_0(x)$$

for all $x \in \overline{E}$. Then for any G-fine division $D = \{(I, x)\}$ of E, we obtain

$$\left| (D) \sum F_{m(x)}(I) - F(E) \right| \leq (D) \sum |F_{m(x)}(I) - F(I)| \quad (3.24)$$

$$\leq (D) \sum |F_{m(x)}(I) - f_{m(x)}(x)\iota(I)|$$

$$+ (D) \sum |f_{m(x)}(x) - f(x)|\iota(I)$$

$$+ (D) \sum |f(x)\iota(I) - F(I)|$$

$$< \sum_{n=1}^{\infty} \frac{\varepsilon}{2^n} + \varepsilon \times \iota(E) + \varepsilon$$

$$= [\iota(E) + 2]\varepsilon.$$

Hence, we have proved the necessity part of the theorem. Conversely, suppose the necessary condition of the theorem is satisfied. Using the same notations as above, we choose an integer $m(x) \geq M(x)$ such that

$$|f_{m(x)}(x) - f(x)| < \varepsilon.$$

Then modify the gauge G if necessary so that $G(x) \subseteq G_{m(x)}(x)$ for each $x \in \overline{E}$. Thus, for any G-fine division $D = \{(I, x)\}$ of E, we get

$$\left| (D) \sum f(x)\iota(I) - F(E) \right| \leq (D) \sum |f(x) - f_{m(x)}(x)|\,\iota(I)$$

$$+ (D) \sum |f_{m(x)}(x)\iota(I) - F_{m(x)}(I)|$$

$$+ \left| (D) \sum F_{m(x)}(I) - F(E) \right|$$

$$< \varepsilon \times \iota(E) + \sum_{n=1}^{\infty} \frac{\varepsilon}{2^n} + \varepsilon$$

$$= [\iota(E) + 2]\varepsilon.$$

Hence, f is $\overline{H}$-integrable to $F(E)$ on E. $\qquad\square$

Remark 3.1 *In the above theorem, it is necessary that the condition holds for all $m(x) \geq M(x)$. However, it is sufficient to have only infinitely many $m(x) \geq M(x)$ in order that f is H-integrable on E with primitive F. We also remark that in the above proof we do not require $F_n(I)$ to converge to $F(I)$ for each subinterval I of E as $n \to \infty$.*

Our next task is to show that the generalised mean convergence theorem (Theorem 3.12) can be deduced from the basic convergence theorem. To this end, we shall prove the following result.

Theorem 3.14 *Let $\{F_n\}$ be a sequence of elementary-set functions. If $\{F_n\}$ satisfies hypothesis (ii) of Theorem 3.12, then the following condition holds. For every $\varepsilon > 0$ there is a function $M(x)$ taking integer values such that for infinitely many $m(x) \geq M(x)$ there is a gauge G on E such that for any G-fine division $D = \{(I, x)\}$ of E, we have*

$$\left| (D) \sum F_{m(x)}(I) - F(E) \right| < \varepsilon.$$

Proof. Let $\varepsilon > 0$ be given. For any integers i and j there exist a gauge $G_{i,j}$ on E and an integer $n(i,j)$ such that for any $G_{i,j}$-fine partial division $D = \{(I, x)\}$ of E with $x \in X_i$, we have

$$\left| (D) \sum [F_n(I) - F(I)] \right| < \frac{\varepsilon}{2^{i+j}}$$

for all $n \geq n(i,j)$. Take $n = n(i,j)$ so that the above inequality holds. We may assume that for each i the sequence $\{F_{n(i+1,j)}\}$ is a subsequence of $\{F_{n(i,j)}\}$. Now consider $F_{n(j)} := F_{n(j,j)}$ in place of F_n and write

$$Y_i = X_i \setminus \bigcup_{k=0}^{i-1} X_k$$

for $i = 1, 2, \ldots$ with X_0 being empty. Put $M(x) = n(i)$ when

$x \in Y_i$. If $m(x)$ takes values in $\{n(j) : j \geq i\}$ when $m(x) \geq M(x) = n(i)$, put

$$G(x) = G_{m(x),j}(x).$$

Note that there are infinitely many such functions $m(x)$. Then for any G-fine division $D = \{(I, x)\}$ of E, we obtain

$$\left| (D) \sum \left[F_{m(x)}(I) - F(I) \right] \right| < \sum_{j=1}^{\infty} \sum_{i=1}^{\infty} \frac{\varepsilon}{2^{i+j}}$$
$$= \varepsilon$$

as desired.

□

So we have proved that the necessary and sufficient condition of the basic convergence theorem holds if hypothesis (ii) of Theorem 3.12 holds. Therefore the generalised mean convergence theorem follows from the basic convergence theorem.

We remark that in the above proof it is necessary to consider i and j where i runs over all Y_i and j runs over all $F_{n(j)}$ as is the case in the proof of Theorem 3.5.

We shall next prove the controlled convergence theorem in a few lemmas. The proof we shall present here is an extension of that given in [23].

Throughout the remainder of this section, given a sequence $\{f_n\}$ of functions and a measurable subset Y of $\overline{E}$, we shall write, for each n,

$$f_{n,Y} = f_n \chi_Y.$$

If $f_{n,Y}$ is H-integrable on E, its primitive will be denoted by $F_{n,Y}$. Given a function f which is H-integrable on E, the notations f_Y and F_Y will be used in a similar manner.

Lemma 3.2 *Let $Y \subseteq \overline{E}$ be a closed set and $f, f_1, f_2, \ldots$ be real-valued functions on $\overline{E}$. Suppose that the following conditions are satisfied.*

(i) $f_{n,Y}(x) \to f_Y(x)$ *almost everywhere in* $\overline{E}$ *as* $n \to \infty$ *where each* $f_{n,Y}$ *is H-integrable on* E *with primitive* $F_{n,Y}$;

(ii) *The sequence* $\{F_{n,Y}\}$ *is* $UAC^\Delta(Y)$.

Then the function f_Y *is H-integrable on* E *with integral value* $F_Y(E)$ *and*

$$\lim_{n\to\infty} F_{n,Y}(E) = F_Y(E).$$

Proof. Since for each n since the function $f_{n,Y}$ is H-integrable on E and its primitive $F_{n,Y}$ is $AC^\Delta(Y)$, by Theorem 3.4, $f_{n,Y}$ is absolutely H-integrable on E and so is Lebesgue integrable on E. Let the function $k_{m,n,Y}$ be given by

$$k_{m,n,Y}(x) = \min\{f_{i,Y}(x) : m \le i \le n\}$$

which is clearly Lebesgue integrable and so is absolutely H-integrable on E. Obviously, for any fixed m, the sequence $\{k_{m,n,Y}\}_{n=1}^\infty$ is monotone. Applying Levi's theorem [17, Theorem 12.22], we infer that

$$k_{m,Y}(x) := \lim_{n\to\infty} k_{m,n,Y}(x)$$

exists almost everywhere in $\overline{E}$ and the function $k_{m,Y}$ is H-integrable on E with integral value given by

$$(H)\int_E k_{m,Y} = \lim_{n\to\infty}\left[(H)\int_E k_{m,n,Y}\right].$$

Since for almost all x, the sequence $\{k_{m,Y}(x)\}$ is monotone in m, by applying Levi's theorem again, we obtain

$$\lim_{m\to\infty} k_{m,Y}(x) = f_Y(x)$$

and the function f_Y is H-integable on E with integral value given by

$$F_Y(E) = (H) \int_E f_Y$$

$$= \lim_{m \to \infty} \left[(H) \int_E k_{m,Y} \right]$$

$$= \lim_{m \to \infty} \lim_{n \to \infty} \left[(H) \int_E k_{m,n,Y} \right]$$

$$= \lim_{m \to \infty} \lim_{n \to \infty} \left[(H) \int_E \min \{ f_{i,Y} : m \le i \le n \} \right]$$

$$\le \lim_{m \to \infty} \lim_{n \to \infty} \min_{m \le i \le n} \left\{ (H) \int_E f_{i,Y} \right\}$$

$$\le \liminf_{m \to \infty} \left[(H) \int_E f_{m,Y} \right].$$

By the same token, we can prove that

$$F_Y(E) \ge \limsup_{m \to \infty} \left[(H) \int_E f_{m,Y} \right].$$

Consequently, we obtain

$$\lim_{n \to \infty} F_{n,Y}(E) = \lim_{n \to \infty} \left[(H) \int_E f_{n,Y} \right]$$

$$= F_Y(E)$$

and the proof is complete.   □

Remark 3.2 *Let $Y \subseteq \overline{E}$ be a closed set. Suppose that hypotheses (i) and (ii) of Lemma 3.2 are satisfied and so f_Y is H-integrable on E with primitive F_Y. Then in view of (3.24) following the arguments in the proof of the basic convergence theorem (Theorem 3.13) would yield the following result. For every $\varepsilon > 0$ there exists an integer-valued function $M(x)$ such that for every integer-valued function $m(x) \ge M(x)$, there is a gauge G on E such that for every G-fine partial division $D = \{(I, x)\}$ of E with $x \in Y$, we have*

$$(D) \sum |F_{m(x),Y}(I) - F_Y(I)| < \varepsilon.$$

Lemma 3.3 *Let f_n, $n = 1, 2, \ldots$, be H-integrable functions on E with primitive F_n and let $Y \subseteq \overline{E}$ be a closed set. If $\{F_n\}$ is $UAC^\triangle(Y)$, then for every $\varepsilon > 0$, there exists a gauge G on E, independent of n, such that for every G-fine partial division $D = \{(I, x)\}$ of E with associated points $x \in Y$, we have, for all $n = 1, 2, \ldots$,*

$$(D) \sum |F_{n,Y}(I) - F_n(I)| < \varepsilon.$$

Proof. Let $\varepsilon > 0$ be given. Since $\{F_n\}$ is $UAC^\triangle(Y)$, we choose a gauge G on E and a positive number η, both independent of n, such that the rest of the condition in the definition of $AC^\triangle(Y)$ holds. Next, by Theorem 3.4, for each n the function $f_{n,Y}$ is absolutely H-integrable on E with primitive $F_{n,Y}$. Consequently, for each fixed n, there exists a gauge G_n, finer than G on E, such that for every G_n-fine partial division $D = \{(I, x)\}$ of E, we have

$$(D) \sum |F_n(I) - f_n(x)\iota(I)| < \varepsilon,$$

and

$$(D) \sum |F_{n,Y}(I) - f_{n,Y}(x)\iota(I)| < \varepsilon.$$

Note that we have applied Henstock's lemma (Proposition 1.8) on f_n and $f_{n,Y}$ which are both H-integrable on E. Now take any G-fine partial division $D = \{(J, y)\}$ of E with associated points $y \in Y$. For each J in D, we construct a G_n-fine division of J and denote the union of all such divisions by D_1. Note that D_1 is a G_n-fine partial division of E. Split $D_1 = \{(I, x)\}$ into D_2 and D_3 so that D_2 contains the interval-point pairs (I, x) with $x \in Y$ and D_3 otherwise. Clearly D_2 is a refinement of D and

$$D_1 \setminus D_2 = D_3.$$

Then we obtain

$$\left|(D)\sum[F_{n,Y}(J)-F_n(J)]\right| = \left|(D_1)\sum[F_{n,Y}(I)-F_n(I)]\right|$$

$$\leq \left|(D_2)\sum[F_{n,Y}(I)-F_n(I)]\right|$$

$$+\left|(D_3)\sum[F_{n,Y}(I)-F_n(I)]\right|$$

$$\leq (D_2)\sum|F_{n,Y}(I)-f_{n,Y}(x)\iota(I)|$$

$$+(D_2)\sum|f_n(x)\iota(I)-F_n(I)|$$

$$+\left|(D_3)\sum F_{n,Y}(I)\right|$$

$$+\left|(D_3)\sum F_n(I)\right|.$$

Modify the gauge G if necessary so that $G(x)\cap Y = \emptyset$ when $x \notin Y$ and that $(D_1\setminus D_2)\sum\iota(I) < \eta$ as in the proof of Theorem 3.4. The former condition implies that $F_{n,Y}(I) = 0$ when $x \notin Y$ and so

$$\left|(D_3)\sum F_{n,Y}(I)\right| = 0.$$

On the other hand, note that $(D_1)\sum\iota(I) = (D)\sum\iota(I)$ and thus

$$(D_1\setminus D_2)\sum\iota(I) = (D_1)\sum\iota(I) - (D_2)\sum\iota(I)$$

$$= (D)\sum\iota(I) - (D_2)\sum\iota(I)$$

$$= (D\setminus D_2)\sum\iota(I).$$

Since $\{F_n\}$ is $UAC^{\triangle}(Y)$, and $(D\setminus D_2)\sum\iota(I) < \eta$ where both D and D_2 are G-fine partial divisions of E with the associated points in Y, and D_2 is a refinement of D, it follows that

$$\left|(D_3)\sum F_n(I)\right| = \left|(D_1\setminus D_2)\sum F_n(I)\right|$$

$$= \left|(D\setminus D_2)\sum F_n(I)\right|$$

$$< \varepsilon.$$

Consequently, we have

$$\left|(D)\sum[F_{n,Y}(J)-F_n(J)]\right|<3\varepsilon.$$

Finally, we split D into D' and D'' such that D' contains the interval-point pairs (J,y) such that

$$F_{n,Y}(J)-F_n(J)\geq 0$$

and D'' otherwise. Repeat the above proof with D replaced by D' and D'', we obtain

$$(D)\sum|F_{n,Y}(J)-F_n(J)|<6\varepsilon$$

as desired.

$\square$

Lemma 3.4 *Let $Y\subseteq\overline{E}$ be a closed set and $f,f_1,f_2,\ldots$ be real-valued functions on $\overline{E}$. Suppose that the following conditions are satisfied.*

(i) *$f_{n,Y}(x)\to f_Y(x)$ almost everywhere in $\overline{E}$ as $n\to\infty$ where each $f_{n,Y}$ is H-integrable on E with primitive $F_{n,Y}$;*
(ii) *The sequence $\{F_{n,Y}\}$ is $UAC^\Delta(Y)$.*

Then the function f_Y is H-integrable on E with primitive F_Y and for every $\varepsilon>0$, there exists an integer N such that for each $n\geq N$ there is a gauge G on E satisfying the condition that for any G-fine division $D=\{(I,x)\}$ of E we have

$$(D)\sum|F_{n,Y}(I)-F_Y(I)|<\varepsilon.$$

Proof. Let $\varepsilon>0$ be given. Since $F_{n,Y}$ is $AC^\Delta(Y)$ uniformly in n, we choose a gauge G on E and a positive number η, both independent of n, such that for any G-fine partial division $D=\{(I,x)\}$ of E with $x\in Y$ satisfying $(D)\sum\iota(I)<\eta$ we have, for all n,

$$(D)\sum|F_{n,Y}(I)|<\varepsilon.$$

By virtue of hypothesis (i), Egoroff's theorem (Theorem 3.7) yields an open set U with $\iota(U) < \eta$ such that for $n, m \geq N$ and $x \notin U$, we have

$$|f_n(x) - f_m(x)| < \varepsilon.$$

Since $f_{n,Y}$ is H-integrable on E, there is a gauge G_n which is finer than G on E such that for any G_n-fine division $D = \{(I, x)\}$ of E, we have

$$(D) \sum |f_{n,Y}(x)\iota(I) - F_{n,Y}(I)| < \frac{\varepsilon}{2^n}.$$

Now for each fixed $m, n \geq N$ and for any G-fine division $D = \{(J, y)\}$ of E we take a division $D_1 = \{(I, x)\}$ of E, which refines D and is both G_n-fine and G_m-fine, and write $D_1 = D_2 \cup D_3$ so that D_2 contains the interval-point pairs (I, x) with $x \notin U$ and D_3 otherwise. We then obtain

$$(D) \sum |F_{n,Y}(J) - F_{m,Y}(J)| \leq (D_1) \sum |F_{n,Y}(I) - F_{m,Y}(I)|$$
$$\leq (D_2) \sum |F_{n,Y}(I) - f_{n,Y}(x)\iota(I)|$$
$$+ (D_2) \sum |F_{m,Y}(I) - f_{m,Y}(x)\iota(I)|$$
$$+ (D_2) \sum |f_{n,Y}(x) - f_{m,Y}(x)|\iota(I)$$
$$+ (D_3) \sum |F_{n,Y}(I)|$$
$$+ (D_3) \sum |F_{m,Y}(I)|$$
$$< \varepsilon(4 + \iota(E)). \tag{3.25}$$

Finally note that the H-integrability of f_Y follows immediately from Lemma 3.2 and so F_Y exists. With this the required result then follows from (3.25).

□

Remark 3.3 *By applying (3.25), we can provide a second proof to Lemma 3.2. This is done as follows. With (3.25), we can find a subsequence $F_{n(j),Y}$ of $F_{n,Y}$ such that for any partial division $D = \{(I, x)\}$ of E we have*

$$\left|(D) \sum \left[F_{n(j),Y}(I) - F_Y(I)\right]\right| < \frac{\varepsilon}{2^j} \quad \text{for } j = 1, 2, \ldots.$$

Then putting $M(\xi) = n(1)$, *for infinitely many* $m(\xi) = n(j) \geq$ $M(\xi)$ *we have*

$$\left| (D) \sum F_{m(\xi),Y}(I) - F_Y(E) \right| < \sum_{j=1}^{\infty} \frac{\varepsilon}{2^j}$$
$$= \varepsilon.$$

It then follows from Theorem 3.13 that f_Y *is H-integrable on* E *with primitive* F_Y.

Lemma 3.5 *Suppose that the following conditions are satisfied.*

(i) $f_n(x) \to f(x)$ *almost everywhere in* $\overline{E}$ *as* $n \to \infty$ *where each* f_n *is H-integrable on* E *with primitive* F_n;
(ii) *The sequence* $\{F_n\}$ *is* $UACG^{\Delta}$ *on* E;
(iii) *There exist an elementary-set function* F *on* E *which is finitely additive over intervals and a gauge* G *on* E *such that for any* G-*fine interval-point pair* (I, x), *we have* $F_n(I) \to F(I)$ *as* $n \to \infty$.

Then the necessary and sufficient condition in the basic convergence theorem (Theorem 3.13) holds.

Proof. Let $\overline{E} = \bigcup_{i=1}^{\infty} X_i$ such that X_i is closed and the sequence $\{F_n\}$ is $UAC^{\Delta}(X_i)$, uniformly in n, for each i. By applying Theorem 3.4 with Y replaced by X_i, we infer that for each i, the function f_{n,X_i} is absolutely H-integrable on E with primitive F_{n,X_i}. In view of Lemma 3.3, we deduce that for each i, the function F_{n,X_i} is also $AC^{\Delta}(X_i)$ uniformly in n. Then by virtue of Lemma 3.4, for every $\varepsilon > 0$ and X_i, writing $Y = X_i$, there exist a positive integer $n = n(i,j)$ and a gauge G_n which is finer than G on E such that for every G_n-fine division $D = \{(I, x)\}$ with $x \in Y$, we have

$$(D) \sum |F_{n,Y}(I) - F_Y(I)| < \frac{\varepsilon}{2^{i+j+1}}.$$

By Lemma 3.3, we may modify the gauge G_n if necessary so that for the same G_n-fine division D above we also have

$$(D) \sum |F_{n,Y}(I) - F_n(I)| < \frac{\varepsilon}{2^{i+j+1}}.$$

Furthermore, in view of hypothesis (iii), for each such D, we may assume that $n = n(i,j)$ is large enough such that

$$(D) \sum |F_n(I) - F(I)| < \frac{\varepsilon}{2^{i+j+1}}.$$

It then follows from the above three inequalities that

$$(D) \sum |F_Y(I) - F(I)| < \frac{\varepsilon}{2^i}.$$

We may also assume that for each $i = 2, 3, \ldots$, the sequence $\{F_{n(i,j)}\}$ is a subsequence of $\{F_{n(i-1,j)}\}$. Now consider $f_{n(j)} := f_{n(j,j)}$ in place of the original sequence f_n. Let

$$Y_1 = X_1,$$

and

$$Y_i = X_i \setminus \bigcup_{k=1}^{i-1} X_k$$

for $i = 2, 3, \ldots$. For each $x \in \overline{E}$, we let $M(x) = n(i)$ when $x \in Y_i$ and let $m(x)$ take values in $\{n(j) : j \geq i\}$ so that $m(x) \geq M(x)$. For each such function $m(x)$, put $G(x) = G_{m(x)}(x)$. Then for any G-fine division $D = \{(I, x)\}$ of E, writing $m = m(x)$ and $Y = X_i$ when $x \in Y_i$, we have

$$|F_m(I) - F(I)| \leq |F_m(I) - F_{m,Y}(I)|$$
$$+ |F_{m,Y}(I) - F_Y(I)|$$
$$+ |F_Y(I) - F(I)|.$$

Consequently, we obtain

$$(D) \sum |F_{m(x)}(I) - F(I)| < 3\varepsilon.$$

Hence the necessary and sufficient condition in Theorem 3.13 is satisfied. $\square$

We have thus proved the controlled convergence theorem for the H-integral as formulated in the following theorem.

Theorem 3.15 (Controlled Convergence Theorem). *Suppose that the following conditions are satisfied.*

(i) $f_n(x) \to f(x)$ *almost everywhere in* $\overline{E}$ *as* $n \to \infty$ *where each* f_n *is* H*-integrable on* E *with primitive* F_n*;*

(ii) *The sequence* $\{F_n\}$ *is* $UACG^\Delta$ *on* E*;*

(iii) *There exist an elementary-set function* F *on* E *which is finitely additive over intervals and a gauge* G *on* E *such that for any* G*-fine interval-point pair* (I, x)*, we have* $F_n(I) \to F(I)$ *as* $n \to \infty$.

Then f *is* H*-integrable on* E *and*

$$\lim_{n \to \infty} (H) \int_E f_n = (H) \int_E f.$$

Remark 3.4 *We may define the* ACG_Δ *and* ACG^Δ *conditions, in Definitions 3.4 and 3.5 respectively, without requiring* X_i *to be closed but only measurable. With this relaxation in the definition of the* ACG^Δ *condition, the assertion of the controlled convergence theorem (Theorem 3.15) still holds. Indeed, the proof may proceed as follows. Suppose the sequence* $\{F_n\}$ *is* $UAC^\Delta(X_i)$ *for each* i *and* $\bigcup_{i=1}^{\infty} X_i = \overline{E}$. *Then by Condition* (∗) *for each* i *there is a closed set* $Y_i \subseteq X_i$ *such that the sequence* $\{F_n\}$ *is* $UAC^\Delta(Y_i)$ *and*

$$Z := \overline{E} \setminus \bigcup_{i=1}^{\infty} Y_i$$

is of measure zero. In view of the $UACG^\triangle$ condition on E, the sequence $\{F_n\}$ satisfies the uniform strong Lusin condition. Then together with Henstock's lemma, we infer that $(H)\int_Z f = 0$. Hence, the proof goes through as before, as is the case in [27]. Note that since Z is of measure zero, f is H-integable on Z and so writing $(H)\int_Z f$ is meaningful.

If a sequence $\{f_n\}$ of functions and a function f satisfy hypotheses (i), (ii) and (iii) of Theorem 3.15, we say that the sequence $\{f_n\}$ is *control-convergent* to f on E. The controlled convergence theorem can then be stated as follows.

If a sequence $\{f_n\}$ of functions is control-convergent to a function f on E, then f is H-integrable on E and

$$(H)\int_E f_n \to (H)\int_E f \text{ as } n \to \infty.$$

Chapter 4

The Radon–Nikodým Theorem
for the H-Integral

The Radon–Nikodým theorem is an important result which is fundamental in measure theory and is usually applied to deal with problems in abstract analysis. It also provides us with the notion of "derivative" of a function in measure spaces where the usual definition of differentiation does not apply. The main theorem we shall prove in this chapter is thus the Radon–Nikodým theorem for the H-integral. What we shall present here is a generalisation of the results we have proved in [28] which are for higher dimensional Euclidean spaces, and those in [36] which are for metric spaces. A generalisation of the Henstock–Kurzweil integral to higher dimensions is the HK-integral (see [21, Chapter 5]) which we shall define in Section 4.3.

The main theorem, namely the Radon–Nikodým theorem for the H-integral, will be proved in Section 4.1. With the aid of the main theorem, we shall characterise the primitive of an H-integrable function in Section 4.2. By modifying the hypothesis of the main theorem, a second version of the theorem will also be given. The corresponding results for the HK-integral and the Henstock–Kurzweil integral which we obtained in [28] will be given in Section 4.3.

4.1   The Main Theorem

The objective of this section is to prove the Radon–Nikodým theorem for the H-integral. Recall that E is an elementary set with a finite measure. We first state without proof the well-known Radon–Nikodým theorem for the Lebesgue integral (for reference see [17, Theorem 19.23]). What we shall state here is not the original statement of the theorem but a form we will use later.

Theorem 4.1 *Let F be a non-negative function, defined on the set of all measurable subsets Y of $\overline{E}$, which is finitely additive over measurable sets and absolutely continuous on E with respect to ι. Then there exists a non-negative function f which is Lebesgue integrable on E such that for any measurable subset Y of $\overline{E}$, we have*

$$F(Y) = (L) \int_Y f. \tag{4.1}$$

We note that, in the above theorem, f is unique in the sense that if g is another non-negative Lebesgue integrable function for which (4.1) holds with f replaced by g, then $f = g$ almost everywhere in $\overline{E}$.

Here, F is absolutely continuous on E with respect to ι if for every $\varepsilon > 0$, there exists $\eta > 0$ such that for every measurable subset Y of $\overline{E}$ satisfying the condition $\iota(Y) < \eta$, we have $|F(Y)| < \varepsilon$, where $|F(Y)|$ denotes the measure of $F(Y)$ on the real line. Note that this definition is standard in measure theory and there should be no confusion with the AC condition defined in Definition 2.2 though they are related in certain sense. The finite additivity of F over measurable sets is defined in the standard manner.

We recall that in Chapter 2 we have established the fact that a measurable function f is Lebesgue integrable on a measurable set $Y \subseteq \overline{E}$ if and only if it is absolutely H-integrable on Y.

Throughout the remainder of this chapter, F shall denote a real-valued elementary-set function which is finitely additive over subintervals of E. Our objective is to define a sequence of non-negative finitely additive measures on E in terms of F so that we can apply Theorem 4.1 to obtain a sequence of Lebesgue integrable functions, and consequently an H-integrable function on E. We begin with a few definitions.

Let $Y \subseteq \overline{E}$ be a measurable set and F be an elementary-set function. We recall Definition 3.4 for the definition of F being $AC_{\Delta}(Y)$. Also recall that if there exist closed sets $X_1, X_2, \ldots$ whose union is $\overline{E}$ such that F is $AC_{\Delta}(X_i)$ for each i, then we say that F is ACG_{Δ} on E. Note that in the definitions below, each X_i is a measurable set instead of a closed set.

Definition 4.1 Let $\{X_i\}$ be a sequence of measurable subsets of $\overline{E}$. An elementary-set function F is said to satisfy the (L)-condition on $\{X_i\}$ if for each fundamental subinterval I_0 of E and every $\varepsilon > 0$ there is a positive integer N such that for any $i \geq N$ there exists a gauge G_i on E satisfying the condition that for any G_i-fine division $D = \{(I, x)\}$ of I_0 we have

$$\left| (D) \sum_{x \notin X_i} F(I) \right| < \varepsilon.$$

Here $(D) \displaystyle\sum_{x \notin X_i}$ sums over all interval-point pairs (I, x) in D with $x \notin X_i$.

Definition 4.2 An elementary-set function F is said to be strongly ACG_{Δ} on E if there exist measurable sets $X_1, X_2, \ldots$ whose union is $\overline{E}$ such that F is $AC_{\Delta}(X_i)$ for each i, and if F satisfies the (L)-condition on $\{X_i\}$.

In the above definitions, we may assume that G_i is a candidate for the gauge G in Definition 3.4 with $Y = X_i$.

Let us begin with an elementary-set function F which is strongly ACG_Δ on E. Then there exist measurable sets $X_1, X_2, \ldots$ whose union is $\overline{E}$ such that for each i and for every $\varepsilon > 0$ the condition in the definition of $AC_\Delta(Y)$ holds with Y, G and η replaced by X_i, G_i and η_i respectively. For each i and for each subinterval J of E, we define

$$K_i(J) = \inf_G \sup_{D_G} \left((D_G) \sum_{x \in X_i} F(I) \right)$$

and

$$|K_i|(J) = \inf_G \sup_{D_G} \left((D_G) \sum_{x \in X_i} |F(I)| \right),$$

where in each of the two cases above the infimum is over all gauges G on E and the supremum is over all G-fine divisions $D_G = \{(I, x)\}$ of J. We proceed to define, for each subinterval J of E,

$$F_i(J) = \inf_P \left((P) \sum K_i(I) \right) \tag{4.2}$$

and

$$|F_i|(J) = \inf_P \left((P) \sum |K_i|(I) \right), \tag{4.3}$$

where in each case the infimum is over all partitions $P = \{I\}$ of J.

Given the function F, we shall call the sequence $\{F_i\}$ of elementary-set functions, where each F_i is defined as in (4.2), the *derived sequence* of F on $\{X_i\}$.

We shall prove that for each i, K_i and $|K_i|$ are finitely super-additive over subintervals of E whereas F_i and $|F_i|$ are finitely additive over subintervals of E.

Let $J = J_1 \cup J_2$ where J_1 and J_2 are disjoint subintervals of E and let a positive integer i be fixed. By the definition of K_i, for every gauge G and for every $\varepsilon > 0$, there exist G-fine divisions D_1 and D_2, of J_1 and J_2 respectively, such that

$$K_i(J_1) + K_i(J_2) < (D_1) \sum_{x \in X_i} F(I) + (D_2) \sum_{x \in X_i} F(I) + \varepsilon$$

$$= (D_1 \cup D_2) \sum_{x \in X_i} F(I) + \varepsilon$$

$$\leq \sup_{D_G} \left((D_G) \sum_{x \in X_i} F(I) \right) + \varepsilon,$$

where $D_1 \cup D_2$ is a G-fine division of $J_1 \cup J_2$. It follows from the arbitrariness of G that

$$K_i(J_1) + K_i(J_2) \leq K_i(J) + \varepsilon.$$

Since ε is arbitrary, we have proved that K_i is finitely superadditive. For a given $\varepsilon > 0$, we can also find partitions P_1 and P_2, of J_1 and J_2 respectively, such that

$$(P_1) \sum K_i(I) + (P_2) \sum K_i(I) < F_i(J_1) + F_i(J_2) + \varepsilon.$$

Since J_1 and J_2 are disjoint, that means

$$(P_1 \cup P_2) \sum K_i(I) < F_i(J_1) + F_i(J_2) + \varepsilon$$

where $P_1 \cup P_2$ is a partition of J. It follows that

$$F_i(J) \leq F_i(J_1) + F_i(J_2).$$

On the other hand, by the superadditivity of K_i and the fact that if $P = \{I_r\}_{r=1}^m$ is a partition of $J_1 \cup J_2$ then

$$P_j := \{I_r \cap J_j\}_{r=1}^m$$

is a partition of J_j for $j = 1, 2$, we can prove that F_i is also finitely superadditive. Indeed, we have

$$F_i(J_1) + F_i(J_2) \leq (P_1) \sum K_i(I) + (P_2) \sum K_i(I)$$

$$\leq \sum_{r=1}^m K_i((I_r \cap J_1) \cup (I_r \cap J_2))$$

$$= (P) \sum K_i(I)$$

and since P is arbitrary, we obtain

$$F_i(J_1) + F_i(J_2) \leq F_i(J).$$

Thus, F_i is finitely additive. Following similar arguments as above, we can show that $|K_i|$ is finitely superadditive whereas $|F_i|$ is finitely additive over subintervals of E.

Since F is ACG_Δ on E and so is $ABRS$ on each X_i, the functions K_i and $|K_i|$ are well defined. Let us prove that K_i is AC on E. The corresponding proof for $|K_i|$ is similar.

For brevity, we shall replace K_i by K, and X_i by Y. Let $\varepsilon > 0$ be given. Since F is $AC_\Delta(Y)$, there exist a gauge G_0 on E and $\eta > 0$ such that for all G_0-fine divisions $D = \{(I, x)\}$ of E with $x \in Y$ satisfying the condition that $(D) \sum \iota(I) < \eta$, we have

$$\left| (D) \sum F(I) \right| < \frac{\varepsilon}{2}.$$

Let $D = \{(J_j, y_j) : j = 1, 2, \ldots, m\}$ be a G_0-fine partial division of E such that $\sum_{j=1}^{m} \iota(J_j) < \eta$. For each $j = 1, 2, \ldots, m$, there exists a gauge G_j which is finer than G_0 such that

$$K(J_j) > \sup_{D_{G_j}} \left((D_{G_j}) \sum_{x \in Y} F(I) \right) - \frac{\varepsilon}{2m}, \qquad (4.4)$$

where the supremum is over all G_j-fine divisions D_{G_j} of J_j. Let G be a gauge on E which is finer than every G_j. Then for each $j = 1, 2, \ldots, m$, there exists a G-fine division D_j of J_j such that

$$K(J_j) \leq \sup_{D_G} \left((D_G) \sum_{x \in Y} F(I) \right)$$
$$< (D_j) \sum_{x \in Y} F(I) + \frac{\varepsilon}{2m}$$

where the supremum is over all G-fine divisions D_G of J_j. On the other hand, since D_j is a G-fine and hence a G_j-fine division of J_j, we have, in view of (4.4),

$$K(J_j) > (D_j) \sum_{x \in Y} F(I) - \frac{\varepsilon}{2m}.$$

Consequently, we obtain

$$(D_j) \sum_{x \in Y} F(I) - \frac{\varepsilon}{2m} < K(J_j) < (D_j) \sum_{x \in Y} F(I) + \frac{\varepsilon}{2m}. \qquad (4.5)$$

Next, let

$$D_0 = \bigcup_{j=1}^{m} D_j.$$

Clearly D_0 is a G_0-fine partial division of E and

$$(D_0) \sum_{x \in Y} \iota(I) \le \sum_{j=1}^{m} \iota(J_j) < \eta$$

which implies that

$$\left| (D_0) \sum_{x \in Y} F(I) \right| < \frac{\varepsilon}{2}.$$

Also note that

$$\sum_{j=1}^{m} \left[(D_j) \sum_{x \in Y} F(I) \right] = (D_0) \sum_{x \in Y} F(I).$$

Now, by summing the terms in (4.5) over $j = 1, 2, \ldots, m$, we obtain

$$\sum_{j=1}^{m} \left[(D_j) \sum_{x \in Y} F(I) \right] - \frac{\varepsilon}{2} < \sum_{j=1}^{m} K(J_j) < \sum_{j=1}^{m} \left[(D_j) \sum_{x \in Y} F(I) \right] + \frac{\varepsilon}{2}$$

which yields

$$\left| \sum_{j=1}^{m} K(J_j) - (D_0) \sum_{x \in Y} F(I) \right| < \frac{\varepsilon}{2}.$$

As a result, we get

$$\left| \sum_{j=1}^{m} K(J_j) \right| \le \left| \sum_{j=1}^{m} K(J_j) - (D_0) \sum_{x \in Y} F(I) \right| + \left| (D_0) \sum_{x \in Y} F(I) \right|$$
$$< \frac{\varepsilon}{2} + \frac{\varepsilon}{2}$$
$$= \varepsilon.$$

In view of Definition 2.2 and the remark thereafter, we have proved that K, which denotes each K_i, is AC on E. Note that whenever $D = \{(J, y)\}$ is a partial division of E, we have

$$(D) \sum F_i(J) = (D) \sum \left[\inf_P \left((P) \sum K_i(I) \right) \right]$$
$$\le (D) \sum K_i(J).$$

It follows that F_i and $|F_i|$ are also AC on E for each i.

For each i and each subinterval I of E, we further define

$$F_i^+(I) = \frac{|F_i|(I) + F_i(I)}{2}$$

and

$$F_i^-(I) = \frac{|F_i|(I) - F_i(I)}{2}.$$

It is easy to see that F_i^+ and F_i^- are well-defined non-negative elementary-set functions such that for each subinterval I of E,

$$F_i(I) = F_i^+(I) - F_i^-(I).$$

For each i, since F_i and $|F_i|$ are finitely additive over subintervals of E and are AC on E, so are F_i^+ and F_i^-.

Let us extend the domains of F_i^+ and F_i^- to all measurable subsets of $\overline{E}$. First, for each $i = 1, 2, \ldots$ and for each measurable subset Y of $\overline{E}$, we define

$$F_i^{++}(Y) = \inf \left\{ \sum_{j=1}^{\infty} F_i^+(I_j) : Y \subseteq \bigcup_{j=1}^{\infty} I_j \right\} \qquad (4.6)$$

and

$$F_i^{--}(Y) = \inf\left\{\sum_{j=1}^{\infty} F_i^{-}(I_j) : Y \subseteq \bigcup_{j=1}^{\infty} I_j\right\}, \qquad (4.7)$$

where the I_j, $j = 1, 2, \ldots$, are subintervals of E. Clearly, the functions F_i^{++} and F_i^{--} agree with F_i^+ and F_i^- respectively on each subinterval of E. Since F_i^+ and F_i^- are AC on E, we can prove that F_i^{++} and F_i^{--} are finitely additive over measurable subsets of $\overline{E}$. Next we note that since $F_i^{++}(Z) = 0 = F_i^{--}(Z)$ if Z is of measure zero, we define

$$F_i^{++}(\overline{E}) = F_i^{++}(E)$$

and

$$F_i^{--}(\overline{E}) = F_i^{--}(E).$$

We shall prove the following lemmas.

Lemma 4.1 *The functions F_i^{++} and F_i^{--} are absolutely continuous on E with respect to ι.*

Proof. Let $\varepsilon > 0$ be given and let $\eta > 0$ be as in the definition of F_i^+ being AC on E. Suppose Y is a measurable subset of $\overline{E}$ such that $\iota(Y) < \eta$. By Condition $(*)$, we can construct a sequence $\{I_k\}_{k=1}^{\infty}$ of disjoint subintervals of E, whose union contains Y except for a set of measure zero, such that $\sum_{k=1}^{\infty} \iota(I_k) < \eta$. Then

$$F_i^{++}(Y) \leq \sum_{k=1}^{\infty} F_i^+(I_k) < \varepsilon.$$

Hence F_i^{++} is absolutely continuous on E with respect to ι. It can be proved similarly that F_i^{--} has the same property. $\square$

Lemma 4.2 *Let I_0 be a fundamental subinterval of E. Suppose that F is strongly ACG_Δ on E and the derived sequence $\{F_i\}$ is as defined in (4.2). Then $F_i(I_0) \to F(I_0)$ as $i \to \infty$.*

Proof.   Let $\varepsilon > 0$ be given and let the sequence of measurable sets $X_1, X_2, \ldots$, the positive integer N as well as the sequence of gauges $G_N, G_{N+1}, \ldots$ on E be as in Definition 4.1. Let $i \geq N$ be fixed. We first choose a partition $P_0 = \{J_1, J_2, \ldots, J_m\}$ of I_0 such that

$$F_i(I_0) - \frac{\varepsilon}{2} < \sum_{k=1}^{m} K_i(J_k) < F_i(I_0) + \frac{\varepsilon}{2}. \qquad (4.8)$$

For each $k = 1, 2, \ldots, m$, there exists a G_i-fine division D_k of J_k such that

$$K_i(J_k) - \frac{\varepsilon}{2m} < (D_k) \sum_{x \in X_i} F(I) < K_i(J_k) + \frac{\varepsilon}{2m}. \qquad (4.9)$$

Next, we define

$$D_0 = \bigcup_{k=1}^{m} D_k.$$

Clearly D_0 is a G_i-fine division of I_0 and summing the terms in (4.9) over $k = 1, 2, \ldots, m$ yields

$$\sum_{k=1}^{m} K_i(J_k) - \frac{\varepsilon}{2} < (D_0) \sum_{x \in X_i} F(I) < \sum_{k=1}^{m} K_i(J_k) + \frac{\varepsilon}{2}.$$

Then together with (4.8), we obtain

$$F_i(I_0) - \varepsilon < (D_0) \sum_{x \in X_i} F(I) < F_i(I_0) + \varepsilon,$$

where

$$(D_0) \sum_{x \in X_i} F(I) = F(I_0) - (D_0) \sum_{x \notin X_i} F(I)$$

in view of the finite additivity of F over subintervals of E. It follows that

$$\left| F(I_0) - F_i(I_0) - (D_0) \sum_{x \notin X_i} F(I) \right| < \varepsilon.$$

By virtue of the (L)-condition, we infer that

$$\left| (D_0) \sum_{x \notin X_i} F(I) \right| < \varepsilon.$$

Consequently, we have

$$|F_i(I_0) - F(I_0)| < 2\varepsilon$$

and the result follows. □

We will need the basic convergence theorem (Theorem 3.13) to prove the Radon–Nikodým theorem for the H-integral. As such, we need to introduce the following concept.

Definition 4.3 Let $F, F_1, F_2, \ldots$ be elementary-set functions. The sequence $\{F_i\}$ is said to satisfy the basic condition with F if for every $\varepsilon > 0$ there is a function $M(x)$ taking integer values such that for infinitely many $m(x) \geq M(x)$ there is a gauge G on E satisfying the condition that for any G-fine division $D = \{(I, x)\}$ of E we have

$$\left| (D) \sum F_{m(x)}(I) - F(E) \right| < \varepsilon.$$

The following result is also essential in proving the main theorem.

Theorem 4.2 *If a function f is H-integrable on E and*

$$(H) \int_{I_0} f = 0$$

for all fundamental subintervals I_0 of E, then $f(x) = 0$ almost everywhere in $\overline{E}$.

Proof. Let $W = \{x \in \overline{E} : f(x) > 0\}$. We shall prove that W is of measure zero. To this end, we define, for each positive integer n,

$$W_n = \{x \in W : 0 < f(x) \leq n\}$$

and $f_n = f\chi_{W_n}$. Let a positive integer n be fixed. By Theorem 2.3, the function f_n is absolutely H-integrable on E and so is Lebesgue integrable on E with integral value given by the Henstock variation

$$A = \inf_{G} \sup_{D_G} \left((D_G) \sum |F_n(I)| \right),$$

where F_n is the primitive of f_n, as in the proof of Theorem 2.2. In view of the hypothesis and by applying Henstock's lemma, we see that $A = 0$. Therefore,

$$(L) \int_{W_n} f = (L) \int_E f_n$$
$$= 0.$$

However, $f(x) > 0$ for all $x \in W_n$. Consequently, W_n is of measure zero. Since

$$W = \bigcup_{n=1}^{\infty} W_n,$$

it follows that W is also of measure zero. Similarly we can show that $\{x \in \overline{E} : f(x) < 0\}$ is of measure zero. This completes the proof.   □

It follows from the above theorem that if f and g are functions which are H-integrable on E such that

$$(H) \int_{I_0} f = (H) \int_{I_0} g$$

for all fundamental subintervals I_0 of E, then $f(x) = g(x)$ almost everywhere in $\overline{E}$.

We are now ready to prove the Radon–Nikodým theorem for the H-integral which is stated in the following theorem.

Theorem 4.3 *Let F be an elementary-set function which is finitely additive over subintervals of E and is strongly ACG_Δ on E such that its derived sequence satisfies with F the basic condition. Then there exists a function f which is H-integrable on E such that*

$$F(E_0) = (H) \int_{E_0} f \qquad (4.10)$$

for all fundamental subsets E_0 of E. Moreover, f is unique in the sense that if g is any function which is H-integrable on E for which (4.10) holds with f replaced by g, then $f = g$ almost everywhere in $\overline{E}$.

Proof. Let $\varepsilon > 0$ be given and let $X_1, X_2, \ldots$, and N be as in Definition 4.1. We may assume that the sequence $\{X_i\}$ is monotone increasing, that is, $X_i \subseteq X_{i+1}$ for $i = 1, 2, \ldots$. We define F_i, $|F_i|$ and subsequently F_i^{++} and F_i^{--} as in (4.2), (4.3), (4.6) and (4.7) respectively. We recall that F_i^{++} and F_i^{--} are non-negative functions which are finitely additive over measurable sets. Furthermore, by Lemma 4.1, the functions F_i^{++} and F_i^{--} are absolutely continuous on E with respect to ι. Hence, by Theorem 4.1, for each i there exist non-negative functions f_i^+ and f_i^- which are Lebesgue integrable on E such that

$$F_i^{++}(Y) = (L) \int_Y f_i^+ \quad \text{and} \quad F_i^{--}(Y) = (L) \int_Y f_i^-$$

for all measurable subsets Y of $\overline{E}$. It follows from the definitions of K_i and $|K_i|$ that whenever $I \cap X_i = \emptyset$, we have

$$K_i(I) = 0 \quad \text{and} \quad |K_i|(I) = 0.$$

Consequently, we have $F_i^{++}(I) = 0$ and $F_i^{--}(I) = 0$ whenever $I \cap X_i = \emptyset$, that is,

$$(L) \int_I f_i^+ = 0 \quad \text{and} \quad (L) \int_I f_i^- = 0$$

for all subintervals I of E such that $I \cap X_i = \emptyset$. Therefore f_i^+ and f_i^- vanish almost everywhere in $\overline{E} \setminus X_i$. Now for each $i = 1, 2, \ldots$, we define

$$f_i(x) = f_i^+(x) - f_i^-(x)$$

for each $x \in \overline{E}$ and then define a function f on $\overline{E}$ given by

$$f(x) = f_i(x)$$

when $x \in X_i$, $i = 1, 2, \ldots$. The function f is well defined due to the uniqueness of f_i on X_i and obviously $f_i(x) \to f(x)$ almost everywhere in $\overline{E}$ as $i \to \infty$. Also note that f_i is Lebesgue integrable on E and so is absolutely H-integrable on E with primitive F_i for each i. Moreover, since the derived sequence of F satisfies with F the basic condition, by Theorem 3.13, the function f is H-integrable on E with primitive F. Now note that for each subinterval I of E, we have

$$F_i(I) = F_i^{++}(I) - F_i^{--}(I)$$

$$= (L) \int_I \left[f_i^+ - f_i^- \right]$$

$$= (L) \int_{I \cap X_i} f$$

where the last equality holds because f_i^+ and f_i^- vanish almost everywhere in $\overline{E} \setminus X_i$. Then by applying Lemma 4.2 and the H-integrability of f, we see that for each fundamental subinterval I of E, we have

$$F(I) = \lim_{i \to \infty} F_i(I)$$

$$= \lim_{i \to \infty} (L) \int_{I \cap X_i} f$$

$$= (H) \int_I f.$$

Since F is finitely additive over subintervals of E, for any fundamental subset E_0 of E, we have $F(E_0) = (H) \int_{E_0} f$ as desired.

The uniqueness of such a function f follows from Theorem 4.2. This completes the proof. $\square$

The function f obtained in Theorem 4.3 shall be called the *Radon–Nikodým derivative*, or briefly *RN-derivative*, of F on E and we say that F is *Radon–Nikodým differentiable*, or briefly *RN-differentiable*, on E. We shall also use the notation $f = D_{RN}F$.

We remark that in view of Theorem 3.13, the derived sequence of the primitive F of an H-integrable function satisfies the basic condition with F. In other words, that the derived sequence of a primitive F satisfies with F the basic condition is a property of the primitive and is therefore not an additional condition imposed on F in Theorem 4.3. We shall prove in the next section that the primitive of an H-integrable function is strongly ACG_Δ on E.

4.2   Descriptive Definition of the H-Integral

The descriptive definition of the Henstock–Kurzweil integral on the real line is well known. More precisely, a function f is Henstock–Kurzweil integrable on a closed interval $[a, b]$ if and only if there exists a function F which is generalised absolutely continuous in the restricted sense on $[a, b]$ such that its derivative $F'(x) = f(x)$ almost everywhere (see [21, Theorem 6.22] or [41]). Thus, the primitive of a Henstock–Kurzweil integrable function on the real line is totally characterised.

A natural generalisation of the Henstock–Kurzweil integral to higher dimensional Euclidean spaces is the HK-integral which we shall define in Section 4.3. We have provided a full characterisation of the primitive of an HK-integrable function in the Euclidean space in [28]. This was not possible previously because the fundamental theorem of calculus was not available for the HK-integral unless some kind of regularity condition is imposed. We have also provided a full characterisation of the primitive of an H-integrable function in metric spaces in [36]. In this section, we shall characterise the primitive of an H-integrable function on measure spaces endowed with locally compact metrizable topologies.

We first prove that the primitive of a function which is H-integrable on E is a strongly ACG_Δ function on E. A similar result for a Henstock-type integral in the Euclidean space has been proved in [27]. Our proof here is an independent one. Also recall that in Proposition 3.1, we prove that the primitive of a function which is H-integrable on E is weakly ACG_Δ on $\overline{E} \setminus Z$ where Z is a set of measure zero. The following result is stronger than Proposition 3.1 and its proof is different from that of the latter.

Theorem 4.4 *If a function f is H-integrable on E, then its primitive F is strongly ACG_Δ on E.*

Proof. By Proposition 1.6, the primitive F of the function f is an elementary-set function which is finitely additive over subintervals of E. Since f is H-integrable on E, by Theorem 3.1 and in view of Theorem 3.2, the function f has a monotone increasing basic sequence $\{X_i\}$ on E such that f is absolutely H-integrable on each X_i and satisfies the (LG)-condition on $\{X_i\}$. For each positive integer i, let

$$f_{X_i} = f\chi_{X_i},$$

where χ_{X_i} denotes, as usual, the characteristic function of X_i. For convenience we write $X_i = Y$. Since f is absolutely H-integrable on Y, it follows from Theorem 2.6 that the primitive F_Y of f_Y is AC on E. Now, by Henstock's lemma, for every

$\varepsilon > 0$ there exists a gauge G on E such that for any G-fine partial division $D = \{(I, x)\}$ of E, we have

$$(D) \sum |F(I) - f(x)\iota(I)| < \varepsilon$$

and for every G-fine McShane partial division $D = \{(I, x)\}$ of E, we have

$$(D) \sum |F_Y(I) - f_Y(x)\iota(I)| < \varepsilon.$$

It follows that F is $ABRS$ on Y. Furthermore, for any G-fine partial division $D = \{(I, x)\}$ of E with $x \in Y$ we have

$$(D) \sum |F(I) - F_Y(I)| \le (D) \sum |F(I) - f(x)\iota(I)|$$
$$+ (D) \sum |F_Y(I) - f_Y(x)\iota(I)|$$
$$< 2\varepsilon.$$

We next choose $\eta > 0$ and modify the gauge G if necessary such that the condition of the definition of F_Y being AC on E holds for the chosen η and the modified G. Then for every G-fine partial division $D = \{(I, x)\}$ with $x \in Y$ such that $(D) \sum \iota(I) < \eta$, we obtain

$$\left| (D) \sum F(I) \right| \le (D) \sum |F(I) - F_Y(I)|$$
$$+ (D) \sum |F_Y(I)|$$
$$< 3\varepsilon.$$

Hence, F is $AC_\Delta(Y)$ and consequently ACG_Δ on E. Finally we note that for any fundamental subinterval I_0 of E and for any G-fine division D of I_0, we have

$$\left| (D) \sum_{x \notin X_i} F(I) \right| \le \left| (D) \sum_{x \notin X_i} f(x)\iota(I) \right|$$
$$+ \left| (D) \sum_{x \notin X_i} [f(x)\iota(I) - F(I)] \right|.$$

Since f satisfies the (LG)-condition on $\{X_i\}$, it follows from the above inequality that F satisfies the (L)-condition. Therefore F is strongly ACG_Δ on E.

□

We remark that in the proof of the above theorem, that F is $AC_\Delta(X_i)$ for each i can also be deduced from Theorem 3.1 and Theorem 2.1.

We further note that $f\chi_{X_i}(x) \to f(x)$ almost everywhere in $\overline{E}$ as $i \to \infty$ and for each i, the function F_i in the derived sequence $\{F_i\}$ of F on $\{X_i\}$ is actually the primitive of $f\chi_{X_i}$ as is seen in the proof of Theorem 4.3. Therefore by Theorem 3.13, the derived sequence $\{F_i\}$ of F satisfies with F the basic condition as pointed out in the previous section.

Conclusively, the primitive of an H-integrable function satisfies all the conditions on F in Theorem 4.3. Therefore a descriptive definition of the H-integral can be given as follows.

Theorem 4.5 *An elementary-set function F is the primitive of a function f which is H-integrable on E if and only if F is a strongly ACG_Δ function on E which is finitely additive over subintervals of E such that its derived sequence satisfies with F the basic condition. Furthermore, $D_{RN}F = f$ almost everywhere in $\overline{E}$.*

Our next task is to provide a second version of the main theorem. This time we shall prove it with the aid of Theorem 3.3 instead of Theorem 3.13. In order to apply Theorem 3.3, we need to prove that the (LG)-condition is satisfied. Note that the (L)-condition involves the elementary-set function F whereas the (LG)-condition involves the point function f. The two conditions are equivalent when f is H-integrable on E with primitive F in view of Henstock's lemma. However, when we do not know whether f is H-integrable, we cannot prove that the

(L)-condition implies the (LG)-condition. We therefore need to impose, in addition, the following (LL)-condition.

Definition 4.4 Let $\{X_k\}$ be a sequence of measurable sets with union $\overline{E}$. A sequence $\{F_k\}$ of elementary-set functions is said to satisfy the (LL)-condition on $\{X_k\}$ if for every $\varepsilon > 0$, there exists a positive integer N such that for each $i \geq N$, there exists a gauge G_i on E satisfying the condition that for every G_i-fine division $D = \{(I, x)\}$ of E, we have

$$\left| \sum_{k=i+1}^{\infty} \left[(D_k) \sum F_k(I) \right] \right| < \varepsilon,$$

where $D_k = \left\{ (I, x) \in D : x \in X_k \setminus \bigcup_{j=1}^{k-1} X_j \right\}$ for $k \geq i + 1$.

The following result is pivotal in accomplishing our next task.

Theorem 4.6 *Let f be a function on $\overline{E}$ and let $\{X_i\}$ be a basic sequence of f on E. For each k, let the primitive of $f\chi_{X_k}$ be F_k. Then f satisfies the (LG)-condition on $\{X_i\}$ if and only if $\{F_k\}$ satisfies the (LL)-condition on $\{X_k\}$.*

Proof. Suppose f satisfies the (LG)-condition on $\{X_i\}$ and let $\varepsilon > 0$ be given. Choose a positive integer N such that for every $i \geq N$, there exists a gauge G_i such that for every G_i-fine division $D = \{(I, x)\}$ of E, we have

$$\left| (D) \sum_{x \notin X_i} f(x)\iota(I) \right| < \frac{\varepsilon}{2}.$$

For each $k \geq N$, since $f\chi_{X_k}$ is H-integrable on E, we may assume that for all G_k-fine partial divisions $D = \{(I, x)\}$ of E, we have

$$(D) \sum \left| f\chi_{X_k}(x)\iota(I) - F_k(I) \right| < \frac{\varepsilon}{2^{k+1}}.$$

Now let $i \geq N$ be fixed. We choose a gauge G_i^* on E such that $G_i^*(x) \subseteq G_k(x)$ if $x \in X_k \setminus \bigcup_{j=1}^{k-1} X_j$ for $k \geq i+1$. Then for every G_i^*-fine division $D = \{(I, x)\}$ of E we let

$$D_k = \left\{ (I, x) \in D : x \in X_k \setminus \bigcup_{j=1}^{k-1} X_j \right\}$$

for $k \geq i+1$. Consequently, we obtain

$$\left| \sum_{k=i+1}^{\infty} \left[(D_k) \sum F_k(I) \right] \right| \leq \left| \sum_{k=i+1}^{\infty} \left[(D_k) \sum [f(x)\iota(I) - F_k(I)] \right] \right|$$

$$+ \left| (D) \sum_{x \notin X_i} f(x)\iota(I) \right|$$

$$< \varepsilon.$$

The converse follows in a similar manner except that this time, we consider instead the inequality

$$\left| (D) \sum_{x \notin X_i} f(x)\iota(I) \right| = \left| \sum_{k=i+1}^{\infty} \left[(D_k) \sum f(x)\iota(I) \right] \right|$$

$$\leq \left| \sum_{k=i+1}^{\infty} \left[(D_k) \sum [f(x)\iota(I) - F_k(I)] \right] \right|$$

$$+ \left| \sum_{k=i+1}^{\infty} \left[(D_k) \sum F_k(I) \right] \right|.$$

This completes the proof.   □

With the above result, we arrive at a second version of the Radon–Nikodým theorem for the H-integral. Its proof follows almost the same way Theorem 4.3 is proved except that we apply Theorem 3.3 instead of Theorem 3.13.

Theorem 4.7 *Let F be an elementary-set function which is finitely additive over subintervals of E and strongly ACG_Δ on E with $\{F_i\}$ being its derived sequence on $\{X_i\}$. If $\{F_i\}$ satisfies the (LL)-condition on $\{X_i\}$, then there exists a function f which is H-integrable on E such that*

$$F(E_0) = (H) \int_{E_0} f \qquad (4.11)$$

for all fundamental subsets E_0 of E. Moreover, f is unique in the sense that, if g is any function which is H-integrable on E for which (4.11) holds with f replaced by g, then $f = g$ almost everywhere in $\overline{E}$.

In view of Theorem 4.6, the primitive F of an H-integrable function has its derived sequence $\{F_i\}$ on $\{X_i\}$ satisfying the (LL)-condition on $\{X_i\}$. Hence we can now state a second descriptive definition of the H-integral.

Theorem 4.8 *An elementary-set function F is the primitive of a function f which is H-integrable on E if and only if F is a strongly ACG_Δ function on E which is finitely additive over subintervals of E such that its derived sequence $\{F_i\}$ on $\{X_i\}$ satisfies the (LL)-condition on $\{X_i\}$. Furthermore, $D_{RN}F = f$ almost everywhere in $\overline{E}$.*

4.3 Henstock Integration in the Euclidean Space

As mentioned previously, a natural extension of the Henstock–Kurzweil integral to the Euclidean space is the HK-integral (see [21, Chapter 5]). We have proved in [28] the Radon–Nikodým theorem for the HK-integral in the Euclidean space and obtained some related results for the HK-integral and the Henstock–Kurzweil integral. In particular, we provided in [28] a characterisation of the HK-integral and that of the Henstock–

Kurzweil integral. The purpose of this section is to present a more detailed account of our results in [28] and translate the terminology we have used thus far into the Euclidean space language.

Let us begin with defining an interval in the n-dimensional Euclidean space $\mathbb{R}^n$. Let I be the set of all points $x = (x_1, \ldots, x_n)$ in $\mathbb{R}^n$ where $a_j \leq x_j \leq b_j$ for $j = 1, 2, \ldots, n$. We write

$$I = [a, b] = [a_1, b_1] \times \cdots \times [a_n, b_n],$$

where $a = (a_1, \ldots, a_n)$ and $b = (b_1, \ldots, b_n)$, and call $I = [a, b]$ an *interval*.

If $J = [\alpha, \beta]$ where $\alpha = (\alpha_1, \ldots, \alpha_n)$ and $\beta = (\beta_1, \ldots, \beta_n)$, any point $\gamma = (\gamma_1, \ldots, \gamma_n)$ with $\gamma_j = \alpha_j$ or β_j is called a *vertex* of J.

The distance between $x = (x_1, \ldots, x_n)$ and $y = (y_1, \ldots, y_n)$ is defined to be

$$\|x - y\| = \left[\sum_{j=1}^{n}(x_j - y_j)^2\right]^{\frac{1}{2}}.$$

An open sphere $B(x, r)$ in $\mathbb{R}^n$ with centre x and radius r is the set of all y in $\mathbb{R}^n$ such that $\|y - x\| < r$.

We shall call a finite union of *non-overlapping intervals* (that is, their interiors are disjoint) an *elementary set*. Note that intervals are themselves elementary sets. Also note that intervals and thus elementary sets here are closed and so if E denotes an elementary set then $E = \overline{E}$. In contrast, intervals and elementary sets in the general setting of measure spaces endowed with locally compact Hausdorff topologies need not be closed.

A partial division D of E is a finite collection of interval-point pairs (I, x) where the intervals I are non-overlapping, and their union is a subset of E. If a partial division D is such that the union of the intervals is E, we call D a *division* of E. We shall write $D = \{(I, x)\}$. As usual, x is called the *associated point* of I.

Let $\mathbb{R}^+$ denote the set of all positive real numbers and let $\delta : E \to \mathbb{R}^+$ be a positive function. A partial division $D = \{(I, x)\}$ of E is said to be δ-*fine* if, for each interval-point pair (I, x), we have $I \subseteq B(x, \delta(x))$ and where x is a vertex of I. Since a division of E is a partial division of E, a δ-fine division is similarly defined.

We also call a finite collection P of non-overlapping intervals I whose union is a subset of E a *partial partition* of E and we write $P = \{I\}$. Again, if a partial partition P is such that the union of the intervals is E, we call P a *partition* of E.

The *volume* of an interval

$$I = [a, b] = [a_1, b_1] \times \cdots \times [a_n, b_n],$$

denoted by $v(I)$, is defined by

$$v(I) = \prod_{j=1}^{n} (b_j - a_j).$$

Note that if two intervals I_1 and I_2 are non-overlapping then $v(I_1 \cap I_2) = 0$.

If a set $Y \subseteq E$ is Lebesgue measurable, then the *Lebesgue measure* of Y, denoted by $\nu(Y)$, is defined by

$$\nu(Y) = \inf \left\{ \sum_{i=1}^{\infty} v(I_i) : Y \subseteq \bigcup_{i=1}^{\infty} I_i \right\},$$

where I_i, $i = 1, 2, \ldots$, are subintervals of E. When $f(x) = g(x)$ for all $x \in E \setminus X$ with $\nu(X) = 0$, we say that $f = g$ almost everywhere in E.

Definition 4.5   A real-valued function f defined on E is said to be HK-integrable on E with integral value $F(E)$ if for every $\varepsilon > 0$, there exists $\delta : E \to \mathbb{R}^+$ such that for any δ-fine division $D = \{(I, x)\}$ of E, we have

$$\left| (D) \sum f(x) v(I) - F(E) \right| < \varepsilon.$$

Here $(D) \sum$ denotes the sum over all interval-point pairs (I, x) in D. As usual, we write

$$(HK) \int_E f \, d\nu = F(E).$$

It is easy to see that this integral is uniquely determined and the set of all HK-integrable functions is closed under addition and scalar multiplication. Furthermore, the integral F as an interval function is finitely additive and Henstock's lemma holds (for a reference, see [21, Chapter 5]). An *interval function* is a real-valued function defined on subintervals of E.

Note that the generalised intervals we define in this book include the intervals in the Euclidean space used for defining the HK-integral as a special case. Furthermore, the H-integral and the HK-integral are equivalent in the Euclidean space. Thus, all the results we have proved in Section 1.3 hold true for the HK-integral.

The definitions of a finitely additive interval function F being AC on E, ACG_Δ on E, and strongly ACG_Δ on E are as described in Definitions 2.2, 3.4, and 4.1 where the gauges G and G_i are replaced by positive functions δ and δ_i respectively, together with other minor adjustments. Following the same arguments in the proofs in Section 4.1 of the results relating the derived sequence of F and the basic condition, we can establish the Radon–Nikodým theorem for the HK-integral, stated as follows, which the author had obtained with Lee in [28, Theorem 10].

Theorem 4.9 *Let F be a finitely additive interval function which is strongly ACG_Δ on E such that its derived sequence satisfies with F the basic condition. Then there exists a HK-integrable function f defined on E such that*

$$F(E_0) = (HK) \int_{E_0} f \, d\nu \qquad (4.12)$$

for all subsets E_0 of E. Moreover, f is unique in the sense that if g is any HK-integrable function on E for which (4.12) holds with f replaced by g, then $f = g$ almost everywhere in E.

Note that following the same argument in the proof of Theorem 4.4 we can prove that the primitive F of an HK-integrable function is a finitely additive and strongly ACG_Δ function on E with its derived sequence satisfying with F the basic condition.

Hence, together with Theorem 4.9, we now have a complete characterisation of the primitive of an HK-integrable function.

Theorem 4.10 *An interval function F is the primitive of a function f which is HK-integrable on E if and only if F is a finitely additive and strongly ACG_Δ function on E such that its derived sequence satisfies with F the basic condition.*

We shall next relate the Radon–Nikodým derivative to differentiable functions in the Euclidean space. To this end, we need the following concept.

Given a point function $\mathcal{F}$, we may define a corresponding finitely additive interval function F, called the *associated interval function*, and conversely. As illustrated in [21, page 128], this can be done as follows.

Suppose that $\mathcal{F}$ is a point function defined on E. Let $I = [\alpha, \beta]$ with $\alpha = (\alpha_1, \ldots, \alpha_n)$ and $\beta = (\beta_1, \ldots, \beta_n)$. Write $\gamma = (\gamma_1, \ldots, \gamma_n)$ where $\gamma_j = \alpha_j$ or β_j and let $n(\gamma)$ denote the number of terms in γ for which $\gamma_j = \alpha_j$. If $\mathcal{F}$ is a function of $x = (x_1, \ldots, x_n)$, that is, $\mathcal{F}(x) = \mathcal{F}(x_1, \ldots, x_n)$, then we define

$$F(I) = \sum_\gamma (-1)^{n(\gamma)} \mathcal{F}(x)$$

where the sum is over all vertices γ.

Conversely, suppose that F is an interval function. Let $I = [a, b]$ and $x \in I$ where $a = (a_1, \ldots, a_n)$, $b = (b_1, \ldots, b_n)$ and $x = (x_1, \ldots, x_n)$. Define $\mathcal{F}(x) = 0$ when $x_j = a_j$ for at least one j, and $\mathcal{F}(x) = F([a, x])$ otherwise.

Next, we say that the point function $\mathcal{F}$ associated with an interval function F is differentiable on E with derivative $\mathcal{F}'$ if for each $x \in E$ and for every $\varepsilon > 0$, there is a positive number $\delta(x)$ such that

$$|F(I) - \mathcal{F}'(x)\nu(I)| < \varepsilon \times \nu(I) \tag{4.13}$$

whenever $I \subseteq B(x, \delta(x))$ and where x is a vertex of I. With this concept, we can prove the following corollary as a simple consequence of Theorem 4.9.

Corollary 4.1 *Let $\mathcal{F}$ be a real-valued point function defined on an elementary set $E \subseteq \mathbb{R}^n$ which is associated with an interval function F. Suppose that $\mathcal{F}$ is differentiable on E with derivative $\mathcal{F}'$. Then $\mathcal{F}' = D_{RN}F$ almost everywhere in E.*

Proof. For each $x \in E$ and for every $\varepsilon > 0$, let $\delta(x) > 0$ be such that (4.13) holds for all intervals $I \subseteq B(x, \delta(x))$, where x is a vertex of I, and we obtain

$$(D) \sum |\mathcal{F}'(x)\nu(I) - F(I)| < \varepsilon \times \nu(E)$$

for each δ-fine division $D = \{(I, x)\}$ of E. Then $\mathcal{F}'$ is HK-integrable on E with primitive F. In fact, for each elementary subset E_0 of E, we have

$$(HK) \int_{E_0} \mathcal{F}' d\nu = F(E_0).$$

The desired result then follows from the uniqueness of the Radon–Nikodým derivative which Theorem 4.9 yields.

□

We shall next deduce from Theorem 4.9 two known results for the Henstock–Kurzweil integral. It is well known that the Henstock–Kurzweil integral is a special case of the HK-integral in $\mathbb{R}^n$ when $n = 1$ (see [21]). In what follows, the function $\mathcal{F}$ is always a point function.

Theorem 4.11 (Fundamental Theorem of Calculus). *Let* $[a, b]$ *be an interval on the real line* $\mathbb{R}$. *If* $\mathcal{F}$ *is a real-valued point function which is differentiable on* $[a, b]$ *with derivative* $\mathcal{F}'$, *then* $\mathcal{F}'$ *is Henstock–Kurzweil integrable on* $[a, b]$ *and*

$$(H) \int_a^b \mathcal{F}' = \mathcal{F}(b) - \mathcal{F}(a).$$

Proof. The result follows by applying Theorem 4.9 for the case when $n = 1$ and Corollary 4.1.

□

Corollary 4.2 *Let* $[a, b]$ *be an interval on the real line. A function* f *defined on* $[a, b]$ *is Henstock–Kurzweil integrable on* $[a, b]$ *with primitive* $\mathcal{F}$ *if and only if* $\mathcal{F}$ *is an* ACG^* *function such that the derivative* $\mathcal{F}'$ *exists and* $\mathcal{F}'(x) = f(x)$ *almost everywhere in* $[a, b]$.

Here $\mathcal{F}$ is ACG^* if $[a, b]$ is the union of closed sets $X_1, X_2, \ldots$ such that, for each i, the following condition holds. For every $\varepsilon > 0$, there exists $\eta > 0$ such that for any partial partition $P = \{[u, v]\}$ of $[a, b]$ with u or v in X_i satisfying the condition that $(P) \sum |v - u| < \eta$, we have

$$(P) \sum |\mathcal{F}(v) - \mathcal{F}(u)| < \varepsilon$$

(for reference, see [21] or [41]). The definition of the ACG_Δ condition in this section can be viewed as a generalisation of the definition of the ACG^* condition to higher dimensions.

Note that the (L)-condition and the basic condition which we impose on the interval function F associated with the point function $\mathcal{F}$ in the n-dimensional Euclidean space are superfluous when $n = 1$ and if $\mathcal{F}$ is ACG^*.

There is another version of the descriptive definition of the Henstock–Kurzweil integral which involves the strong Lusin condition. Though this is an isolated point, we state it here for interest.

A function f is Henstock–Kurzweil integrable on $[a, b]$ if and only if there is a continuous function $\mathcal{F}$ such that $\mathcal{F}'(x) = f(x)$ almost everywhere in $[a, b]$ and $\mathcal{F}$ satisfies the strong Lusin condition. In this setting a function $\mathcal{F}$ is said to satisfy the strong Lusin condition on $[a, b]$ if for every set $S \subseteq [a, b]$ of measure zero and for every $\varepsilon > 0$, there exists $\delta(\xi) > 0$ for $\xi \in S$ such that for any δ-fine partial division D of interval-point pairs $\{([u, v], \xi)\}$ with $\xi \in S$, we have

$$(D) \sum |\mathcal{F}(v) - \mathcal{F}(u)| < \varepsilon.$$

As is seen in Corollary 4.11, on the real line we have the fundamental theorem of calculus which links the primitive with the integrable function. However, no such simple theorem exists even for the Euclidean space unless some kind of regularity condition is imposed [39]. In measure theory, we have the Radon–Nikodým theorem which is in a sense a version of the fundamental theorem. We contributed our modest effort in proving the Radon–Nikodým theorem for the HK-integral in the Euclidean space in [28] and extending the result to the H-integral on metric spaces in [36]. In this book we extend the result further to the present setting.

Chapter 5

Harnack Extension and Convergence Theorems for the H-Integral

In [24], Lee extends the Harnack extension, an important property of the Henstock–Kurzweil integral on the real line, to the HK integral in the Euclidean space and recovers the proof by means of category argument. Note that in the classical theory of the Denjoy–Perron integral, proving convergence theorems by means of the category argument in which a key step is the use of Harnack extension is a standard approach (for reference, see [21, page 47] or [41, page 253]).

We shall show in this chapter that Harnack extension is valid for the H-integral on measure spaces endowed with locally compact metric topologies, and apply the category argument in proving the Harnack convergence theorem with which we proceed to prove an improved version of the controlled convergence theorem for the H-integral.

5.1 The H-Integral on Metric Spaces

As we shall prove the results in this chapter for the H-integral on measure spaces endowed with locally compact metric topologies, which is, as we have mentioned in the previous chapters, a special case of the general setting of measure spaces endowed with locally compact Hausdorff topologies, we shall first revisit the key ideas for defining the H-integral and restate them in the language of metric space whenever necessary. This, we hope, could be useful for readers who are interested in understanding the H-integral specifically in the metric space setting.

Let (X, d) be a metric space with a locally compact metric topology $\mathcal{T}$ induced by the metric d on X and let (X, Ω, ι) be a measure space such that $\mathcal{T} \subseteq \Omega$. The measure ι is non-negative and countably additive.

Let $\mathcal{T}_1$ be the set of all *d-open balls* or simply *open balls*, that is, sets of the form $\{y \in X : d(x, y) < r\}$, where $x \in X$ and $r > 0$, which are denoted by $B(x, r)$. The point x and the positive real number r are called the *centre* and the *radius* of the open ball $B(x, r)$ respectively. We shall also call the closure of an open ball a *closed ball*. It is assumed that $\iota(B) > 0$ and $\iota(B) = \iota(\overline{B})$ for all open balls B, where $\overline{B}$ denotes, as usual, the closure of B.

We then define

$$\mathcal{I}_0^* = \left\{ \overline{B_1} \setminus \overline{B_2} : B_1, B_2 \in \mathcal{T}_1 \text{ where } B_1 \nsubseteq B_2 \text{ and } B_2 \nsubseteq B_1 \right\},$$

$$\mathcal{I}_1^* = \left\{ \bigcap_{i \in \Lambda} X_i \neq \emptyset : X_i \in \mathcal{I}_0^* \text{ and } \Lambda \text{ is a finite index set} \right\}.$$

Note that $\mathcal{I}_0^*$ and $\mathcal{I}_1^*$ are the respective metric space analogues of $\mathcal{I}_0$ and $\mathcal{I}_1$ defined on page 6. As described in Example 1.2, members of $\mathcal{I}_0^*$ are either closed balls or scalloped balls. A *generalised interval* is defined to be a member of $\mathcal{I}_1^*$ which is typically a finite intersection of a combination of closed balls and scalloped balls. All members I of $\mathcal{I}_1^*$ are relatively compact (that is, $\overline{I}$ is compact) though not necessarily closed or compact, because $\mathcal{T}$ is locally compact. Also note that $\iota(I) = \iota(\overline{I})$ for each interval I.

We define elementary sets, partial divisions, divisions, refinements, and associated points in the metric space setting exactly the same way we define them in Section 1.1. Next, let $\delta : \overline{E} \to \mathbb{R}^+$ be a positive function. We also call δ a *gauge* on E. As in the case of a gauge G in the general setting, we need to consider gauges defined on $\overline{E}$ and not just E because for each interval-point pair (I, x) in a partial division, the associated point x belongs to $\overline{I}$ and not just I.

Let a gauge δ on E be given. An interval-point pair (I, x) is *δ-fine* if $I \subseteq B(x, \delta(x))$. A partial division $\{(I_i, x_i) : i = 1, 2, \ldots, n\}$ of E is *δ-fine* if (I_i, x_i) is δ-fine for each $i = 1, 2, \ldots, n$. Since divisions are themselves partial divisions, δ-fine divisions of E are similarly defined. The existence of δ-fine divisions has been proved by the author and Lee in [37].

A gauge δ_1 is said to be *finer* than a gauge δ_2 on E if for each $x \in \overline{E}$ we have $\delta_1(x) \leq \delta_2(x)$. It is easy to see that given two gauges δ_1 and δ_2 on E, there is always a gauge δ which is finer than both δ_1 and δ_2; we could simply define $\delta(x) = \min(\delta_1(x), \delta_2(x))$ for each $x \in \overline{E}$. Consequently, if D is a δ-fine division of E, then D is both δ_1-fine and δ_2-fine.

The metric space version of the H-integral is similar to that in the general setting, that is, a function f is said to be *H-integrable* on E to a real number A if for every $\varepsilon > 0$, there exists a gauge δ on E such that for every δ-fine division $D = \{(I, x)\}$ of E, we have

$$\left| (D) \sum f(x) \iota(I) - A \right| < \varepsilon.$$

We write $(H) \int_E f = A$. The H-integrability of f on any elementary subset of E is similarly defined.

Note that by definition, if f is H-integrable on E it is also H-integrable on $\overline{E}$, and vice versa. All results obtained for the H-integral in the previous chapters hold for the present setting. In particular, the H-integral is uniquely determined, and closed under addition, scalar multiplication, and monotone convergence. Furthermore, the Cauchy criterion of integrability and Henstock's lemma also hold. If both f and $|f|$ are H-integrable on E, then we say that f is absolutely H-integrable on E.

As we have pointed out in Chapter 2, we can extend the domain of H-integrability to measurable sets. Given a function f on $\overline{E}$ and for every measurable subset W of $\overline{E}$, the function f_W is given by $f_W(x) = f(x)$ if $x \in W$ and 0 otherwise. We can then say that f is H-integrable on W to a real number A if f_W is H-integrable on E to the number A and write $(H) \int_W f = A$. If f_W is absolutely H-integrable on E, we say that f is absolutely H-integrable on W and it is then meaningful to write $(H) \int_W |f|$. Recall that by definition, if f is H-integrable on W to the value A, then

$$(H) \int_W f = (H) \int_E f_W$$
$$= A.$$

If F is the primitive of f on E, we shall write $F(W) = A$.

5.2   Harnack Extension for the H-Integral

Harnack extension on the real line states that if a function f is Henstock–Kurzweil integrable on a closed subset X of an interval $[a, b]$ and on each interval $[a_k, b_k]$ where $(a, b) \setminus X = \bigcup_{k=1}^{\infty} (a_k, b_k)$, and if

$$\sum_{k=1}^{\infty} \omega \left(F; [a_k, b_k] \right) < \infty,$$

then f is Henstock–Kurzweil integrable on $[a, b]$. Here F denotes the primitive of f on $[a_k, b_k]$ and $\omega \left(F; [a_k, b_k] \right)$ is the *oscillation* of F over $[a_k, b_k]$, that is,

$$\omega(F; [a_k, b_k]) = \sup \{|F(x) - F(y)| : x, y \in [a_k, b_k]\}.$$

To extend Harnack extension beyond the real line, we need to reformulate the condition involving oscillation of the primitive function by introducing a concept called nonabsolute set as employed by Lee in [24] which is for the Euclidean space.

To motivate the definition, we consider the situation where F is a continuous function on $[a, b]$ and U is an open subset of $[a, b]$ where $U = \bigcup_{k=1}^{\infty} (a_k, b_k)$. Suppose that there is $M > 0$ such that for any disjoint intervals $I_1, I_2, \ldots, I_m$ in U with each connected component (a_k, b_k) of U containing at most one I_i only, we have $\sum_{i=1}^{m} |F(I_i)| \leq M$. Then we obtain

$$\sum_{k=1}^{\infty} \omega(F; [a_k, b_k]) \leq M.$$

However, if we consider any non-overlapping intervals $I_1, I_2, \ldots, I_m$ in U without restricting each connected component (a_k, b_k) of U to contain at most one I_i, then we will obtain

$$\sum_{k=1}^{\infty} V(F; [a_k, b_k]) \leq M$$

where V denotes the *total variation* of F on $[a_k, b_k]$, that is,

$$V(F; [a_k, b_k]) = \sup_{P} \left((P) \sum |F(v) - F(u)| \right)$$

where the supremum is over all partitions $P = \{(u, v)\}$ of E. The case involving the oscillation will give rise to a nonabsolute integral while that involving the total variation will yield an absolute one. In formulating the Harnack extension for the H-integral, it is therefore the former case that we want and the union of the intervals $I_1, I_2, \ldots, I_m$ thus considered is what we call a nonabsolute set, so named to reflect its close link to non-absolute integrals. We shall make precise the notations in the following definitions.

Definition 5.1 Let E_0 be an elementary subset of E, Y be a closed subset of E_0 and δ be a gauge on E. A δ-fine cover of Y is the union of the intervals $I_1, I_2, \ldots, I_n$ in a δ-fine partial division $\{(I_i, x_i) : i = 1, 2, \ldots, n\}$ of E with associated points $x_i \in Y$ such that the union contains Y.

Definition 5.2 Let E_0 be an elementary subset of E and U be an open subset of E_0. An elementary set E_1 is called a nonabsolute subset of U if there exists a gauge δ on E such that E_1 is the complement of a δ-fine cover of $E_0 \setminus U$ relative to E_0. We say that E_1 is a nonabsolute subset of U involving δ with respect to E_0. Where there is no need to specify the gauge δ and the elementary set E_0, we simply call all such sets nonabsolute subsets of U.

Let $Y = E_0 \setminus U$ where E_0 and U are as in Definition 5.2. Note that since a δ-fine cover of Y is the union of the intervals of a partial division of E, a nonabsolute subset E_1 of $E_0 \setminus Y$, being the complement of a δ-fine cover of Y relative to E_0, is a finite union of disjoint intervals and is thus an elementary set. Also note that since U is open given a gauge δ on E, by modifying δ if necessary such that $B(x, \delta(x)) \cap Y = \emptyset$ if $x \notin Y$, we can ensure that every δ-fine division of E_0 has a subset of which the union of the intervals is a δ-fine cover of Y.

Let us understand the concept of a nonabsolute subset in the context of the real line. Note that it is possible that a δ-fine cover of $Y = E_0 \setminus U$ may be a δ-fine division of E_0. In the following example we shall consider only the case in which the complement of a δ-fine cover of Y is non-empty.

Example 5.1   Consider the case where $X = \mathbb{R}$ and E_0 is a closed bounded interval $[a, b]$. Let Y be a closed subset of $[a, b]$. Then $(a, b) \setminus Y$ is an open set and thus can be written as the union of a countable number of pairwise disjoint intervals $(a_1, b_1), (a_2, b_2), \ldots$. Given $\delta(x) > 0$, let $D = \{(I_i, x_i) : i = 1, 2, \ldots, n\}$ be a δ-fine partial division of E_0 with $x_i \in Y$ such that

$$C := \bigcup_{i=1}^{n} I_i \supseteq Y.$$

Then by definition, C is a δ-fine cover of Y and $[a, b] \setminus C$ is a nonabsolute subset of $[a, b] \setminus Y$. Note that $[a, b] \setminus C$ consists of a finite number of connected bounded intervals $[u_1, v_1], [u_2, v_2], \ldots, [u_m, v_m]$ which we assume to be closed for convenience. Obviously, all except for a finite number of the intervals (a_i, b_i), say $(a_{l_1}, b_{l_1}), (a_{l_2}, b_{l_2}), \ldots, (a_{l_p}, b_{l_p})$, are contained in C. It is clear that each component interval $[u_k, v_k]$ of $[a, b] \setminus C$ lies inside $(a_{l_{q(k)}}, b_{l_{q(k)}})$ for some $1 \leq q(k) \leq p$. Also, each (a_{l_q}, b_{l_q}) contains one and only one of the $[u_k, v_k]$.

We need the following lemma in order to prove Harnack extension for the H-integral.

Lemma 5.1   *Let Y be a closed subset of E. There exist elementary sets $E_1, E_2, \ldots$ such that $E_i \subseteq E \setminus Y$ for each $i = 1, 2, \ldots,$ and*

$$\iota\left((E \setminus Y) \setminus \bigcup_{i=1}^{\infty} E_i\right) = 0.$$

Proof.   By Condition $(*)$, for each $i = 1, 2, \ldots,$ we choose an open set U_i such that $Y \subseteq U_i$ and

$$\iota(U_i \setminus Y) < \frac{1}{i}$$

and define a gauge δ_i such that $B(x, \delta_i(x)) \cap Y = \emptyset$ when $x \in \overline{E} \setminus Y$ and $B(x, \delta_i(x)) \subseteq U_i$ if $x \in Y$. We may assume that

δ_{i+1} is finer than δ_i for each $i = 1, 2, \ldots$. Let $D_i = \{(I, x)\}$ be a δ_i-fine division of E and decompose D_i into $D_i^{(1)}$ and $D_i^{(2)}$ where $D_i^{(1)}$ contains all interval-point pairs (I, x) in D such that the associated points x belong to $\overline{E} \setminus Y$ and $D_i^{(2)}$ contains all the remaining interval-point pairs in D. For each $i = 1, 2, \ldots$, we define

$$E_i = \bigcup_{(I,x) \in D_i^{(1)}} I \quad \text{and} \quad C_i = \bigcup_{(I,x) \in D_i^{(2)}} I$$

and note that $E_i = E \setminus C_i$ and $Y \subseteq C_i \subseteq U_i$. Consequently, for each $i = 1, 2, \ldots$, we have $E_i \subseteq E \setminus Y$ and

$$0 \leq \iota \left((E \setminus Y) \setminus \bigcup_{i=1}^{\infty} E_i \right)$$
$$\leq \iota ((E \setminus Y) \setminus E_i)$$
$$= \iota (C_i \setminus Y)$$
$$\leq \iota (U_i \setminus Y)$$
$$< \frac{1}{i}$$

and hence $\iota \left((E \setminus Y) \setminus \bigcup_{i=1}^{\infty} E_i \right) = 0$ as desired.

$\square$

Note that in the above proof, each C_i is a δ-fine cover of Y and each E_i is a nonabsolute subset of $E \setminus Y$.

We are now ready to prove Harnack extension for the H-integral as stated below. As usual, f_Y denotes the function on $\overline{E}$ such that $f_Y(x) = f(x)$ when $x \in Y$ and 0 otherwise.

Theorem 5.1 *Let f be a function on $\overline{E}$ and Y be a closed subset of E. Suppose that the following conditions are satisfied.*

(i) *f is H-integrable on Y;*
(ii) *f is H-integrable on every elementary subset of $E \setminus Y$;*

(iii) *There is a real number A satisfying the condition that for every $\varepsilon > 0$ there exists a gauge δ on E such that for any nonabsolute subset E_1 of $E \setminus Y$ involving δ, we have*

$$\left| (H) \int_{E_1} f - A \right| < \varepsilon.$$

Then f is H-integrable on E and

$$(H) \int_E f = (H) \int_Y f + A.$$

Proof. Let $g = f - f_Y$. By hypothesis (i), the function f is H-integrable on Y and so the function f_Y is H-integrable on E. Thus, it suffices to show that g is H-integrable on E with the integral A. Since Y is a closed subset of E, by Lemma 5.1, there exist elementary sets $E_1, E_2, \ldots$ such that for each $i = 1, 2, \ldots,$ $E_i \subseteq E \setminus Y$, and

$$\iota \left((E \setminus Y) \setminus \bigcup_{i=1}^{\infty} E_i \right) = 0.$$

Without loss of generality, we may assume that for each $i = 1, 2, \ldots,$ we have $E_i \subseteq E_{i+1}$. Let $\varepsilon > 0$ be given. By hypothesis (ii), the function f is H-integrable on E_i for each i. Apply Henstock's lemma and choose for each i a gauge δ_i on E_i such that for any δ_i-fine partial division $D = \{(I, x)\}$ of E_i we have

$$(D) \sum \left| f(x)\iota(I) - (H) \int_I f \right| < \frac{\varepsilon}{2^i}.$$

We may assume that δ_{i+1} is finer than δ_i for each i. Next, we define a gauge δ^* on E such that when $x \in Y$, we have $\delta^*(x) \leq \delta(x)$, where δ is the gauge on E given in hypothesis (iii), and when $x \in (E_i \setminus E_{i-1})^o$, where $i = 1, 2, \ldots$ and $E_0 = \emptyset$, we have $\delta^*(x) \leq \delta_i(x)$. Since for each i, the set $(E_i \setminus E_{i-1})^o$ is open, we may modify δ^* if necessary such that whenever $x \in (E_i \setminus E_{i-1})^o$, we have $B(x, \delta^*(x)) \subseteq (E_i \setminus E_{i-1})^o$. Let

$$Z = \left((\overline{E} \setminus Y) \setminus \bigcup_{i=1}^{\infty} E_i \right) \bigcup \left(\bigcup_{i=1}^{\infty} \partial E_i \right)$$

Since each E_i is an elementary set, we have $\iota(\partial E_i) = 0$ for each i. Likewise, $\iota(\partial E) = 0$. Hence Z is a set of measure zero and so by virtue of Proposition 1.3, we may assume that $f(x) = 0$ for $x \in Z$. Also note that every δ^*-fine division of E contains a δ^*-fine cover of Y. Now take any δ^*-fine division $D = \{(I, x)\}$ of E and decompose it into D_1 and D_2 where $D_1 = \{(I, x) \in D : x \in Y\}$ and $D_2 = \{(I, x) \in D : x \notin Y\}$. Clearly,

$$(D_1) \sum g(x)\iota(I) = 0$$

and

$$(D_2) \sum g(x)\iota(I) = (D_2) \sum f(x)\iota(I),$$

and thus

$$\left|(D) \sum g(x)\iota(I) - A\right| = \left|(D_2) \sum f(x)\iota(I) - A\right|.$$

Let E^* be the union of all intervals I such that $(I, x) \in D_2$. Obviously, E^* is a nonabsolute subset of $E \setminus Y$ involving δ^*. Since δ^* is finer than δ, by hypothesis (iii), we have

$$\left|(H) \int_{E^*} f - A\right| < \varepsilon.$$

Finally, for each $i = 1, 2, \ldots$, let $D^{(i)}$ denote the subset of D_2 for which the associated points $x \in (E_i \setminus E_{i-1})^o$. Note that only a finite number of the partial divisions $D^{(i)}$ are non-empty. Clearly,

$$(D_2) \sum f(x)\iota(I) = \sum_{i=1}^{\infty} \left[(D^{(i)}) \sum f(x)\iota(I)\right]$$

and

$$\sum_{i=1}^{\infty} \left[(D^{(i)}) \sum (H) \int_I f\right] = (H) \int_{E^*} f.$$

It follows that

$$\left|(D)\sum g(x)\iota(I) - A\right| \le \sum_{i=1}^{\infty}\left[(D^{(i)})\sum\left|f(x)\iota(I) - (H)\int_I f\right|\right]$$

$$+ \left|(H)\int_{E^*} f - A\right|$$

$$< \sum_{i=1}^{\infty}\frac{\varepsilon}{2^i} + \varepsilon$$

$$= 2\varepsilon.$$

Hence the proof is complete.

□

Corollary 5.1   *Theorem 5.1 holds true with $Y = \partial E$, the boundary of E.*

This is known as the Cauchy extension of the H-integral which is a special case of Harnack extension. In fact, Theorem 5.1 should be called, more appropriately, the Cauchy–Harnack extension. Note that in the case of the real line, the Cauchy extension for the Henstock–Kurzweil integral is also a special case of the Harnack extension where $Y = \{a\}$ (for reference, see [21, p. 47]). We remark that f is H-integrable on E if and only if f is H-integrable on $\overline{E}$. This follows from the fact that $\iota(E) = \iota(\overline{E})$ and that if two functions are equal almost everywhere on E, then if one is H-integrable on E so is the other (see [35]). So the validity of the Cauchy extension for the H-integral is implicit in the definition of the H-integral.

5.3   The Category Argument

In this section, we shall prove a convergence theorem, which we shall call the Harnack convergence theorem, by means of the category argument and by applying the Cauchy–Harnack extension of the H-integral. We will first prove the dominated convergence theorem with which we can prove the mean convergence theorem. With the aid of the Harnack convergence theorem, we shall prove an improved version of the controlled convergence theorem for the H-integral in Section 5.4.

In proving the Harnack convergence theorem using the category argument, we will need to apply the well-known Baire category theorem, also known as Baire's theorem or the category theorem. A general form of the theorem states that if Y is a Baire space and if Y is a countable union of closed sets, then one of the closed sets must have a non-empty interior. A *Baire space* is a topological space such that every intersection of a countable collection of open dense sets in the space is also dense. A subset W of a topological space Y is *dense* in Y if for any point x in Y, any neighbourhood of x contains at least one point from W. Here we shall only prove the special case of the theorem when Y is compact.

Theorem 5.2 (Baire Category Theorem). *Let $Y \subseteq \overline{E}$ be a closed set and suppose that*

$$Y = \bigcup_{i=1}^{\infty} Y_i$$

where each Y_i is closed. Then there exists an integer m such that $Y_m^\circ \neq \emptyset$. Here Y_m° denotes, as usual, the interior of Y_m.

Proof. First note that since Y is a closed subset of $\overline{E}$ where the latter is compact, Y as a subspace of $\overline{E}$ is also compact and so is complete. In other words, every Cauchy sequence in Y converges. We shall prove the theorem by contradiction. Suppose no Y_i has a non-empty interior. Then no Y_i contains a non-empty open set and thus no Y_i equals Y. In particular, $Y_1 \neq Y$. Then $Y \setminus Y_1$ is an open set in Y which must contain an open ball $B_1 := B(x_1, \varepsilon_1)$ with $0 < \varepsilon_1 < \frac{1}{2}$. The set Y_2 does not contain the open ball $B\left(x_1, \dfrac{\varepsilon_1}{2}\right)$. Hence the non-empty open set

$(Y \setminus Y_2) \cap B\left(x_1, \dfrac{\varepsilon_1}{2}\right)$ in Y contains an open ball $B_2 := B(x_2, \varepsilon_2)$

with $0 < \varepsilon_2 < \dfrac{1}{4}$. Continuing this process inductively, we obtain a sequence

$$B_k := B(x_k, \varepsilon_k)$$

of open balls such that, for each positive integer k, we have $0 < \varepsilon_k < \dfrac{1}{2^k}$ and

$$B_{k+1} \subseteq B\left(x_k, \frac{\varepsilon_k}{2}\right),$$

as well as $B_k \cap Y_k = \emptyset$. We may assume that the sequence $\{Y_k\}$ is infinite for if it is finite then we are done. Since for $n < m$ we have

$$d(x_n, x_m) \le \sum_{k=n}^{m-1} d(x_k, x_{k+1})$$

$$< \sum_{k=n}^{m-1} \frac{1}{2^{k+1}}$$

$$< \frac{1}{2^n},$$

the centres x_k of the balls B_k form a Cauchy sequence, and so converge to a point x_0 in Y. Since for all $m > n$,

$$d(x_n, x_0) \le d(x_n, x_m) + d(x_m, x_0)$$

$$< \frac{\varepsilon_n}{2} + d(x_m, x_0)$$

and

$$\lim_{m \to \infty} d(x_m, x_0) = 0,$$

it follows that $d(x_n, x_0) \le \dfrac{\varepsilon_n}{2}$ which means that $x_0 \in B_n$ for every $n \ge 1$. Thus, the point x_0 is in none of the sets Y_n, and so is not in their union which is Y. But this is a contradiction because x_0 belongs to Y. Therefore at least one of the closed sets Y_k contains a non-empty open set, that is, there exists an integer m such that Y_m has a non-empty interior. This completes the proof.   □

The above theorem was first proved by W. F. Osgood in 1897 for the real line, and independently by R. Baire in 1899 for the n-dimensional Euclidean space. We will later see that in our setting, if Y_m is a closed set which has a non-empty interior, then it contains an interval which is a closed set.

We shall also need the following definitions.

Definition 5.3 Let E_0 be an elementary subset of E and Y be a closed subset of E_0. A sequence $\{f_n\}$ of functions is said to satisfy the uniform Harnack condition on Y with respect to E_0 if each f_n is H-integrable on every measurable subset W of Y and every elementary subset E_1 of $E_0 \setminus Y$, and there exists a sequence $\{A_n\}$ of real numbers such that the following condition is satisfied. For every $\varepsilon > 0$ there exists a gauge δ on E_0, independent of n, such that for any nonabsolute subset E_1 of $E_0 \setminus Y$ involving δ, we have

$$\left| (H) \int_{E_1} f_n - A_n \right| < \varepsilon \qquad (5.1)$$

for all n. When $f_n = f$ and $A_n = A$ for all n, then f is said to satisfy the Harnack condition on Y with respect to E_0.

It follows from Theorem 5.1 that if a function f satisfies the Harnack condition on Y with respect to E_0, then f is H-integrable on E_0.

Definition 5.4 Let $\{f_n\}$ be a sequence of functions on $\overline{E}$ and Y be a closed subset of $\overline{E}$. We say that $\{f_n\}$ satisfies the uniform (AC)-condition on Y if each f_n is H-integrable on every measurable set $W \subseteq Y$ and for every $\varepsilon > 0$ there exist $\eta > 0$, independent of n, and a positive integer N such that whenever $W \subseteq Y$ and $\iota(W) < \eta$, we have

$$(H) \int_W |f_n| < \varepsilon \tag{5.2}$$

for all $n \geq N$. When $f_n = f$ for all n, then f is said to satisfy the (AC)-condition on Y.

Remark 5.1 *Note that in the above definition, condition (5.2) can be replaced with the condition* "$\left|(H) \int_W f_n\right| < \varepsilon$" *as these two conditions are equivalent. Indeed, if the latter condition holds, then for every measurable subset W of Y satisfying $\iota(W) < \eta$, we can express W as the disjoint union of W_1 and W_2 where W_1 contains all $x \in W$ such that $f(x) > 0$ and W_2 otherwise, and then obtain*

$$(H) \int_W |f_n| \leq \left|(H) \int_{W_1} f_n\right| + \left|(H) \int_{W_2} f_n\right|$$
$$< 2\varepsilon.$$

The converse is obvious as

$$\left|(H) \int_W f_n\right| \leq (H) \int_W |f_n|.$$

Also note that by Theorem 2.4, if a function f is H-integrable on E and its primitive is AC on E, then f satisfies the (AC)-condition on Y.

We shall next prove the dominated convergence theorem by first proving two lemmas. We shall extend the proof in [21, page 14] which is for the real line to the present setting. In the following lemmas, the functions $\max(f_1, f_2)$ and $\min(f_1, f_2)$ are given by

$$\max(f_1, f_2)(x) = \max(f_1(x), f_2(x))$$

and

$$\min(f_1, f_2)(x) = \min(f_1(x), f_2(x))$$

for each $x \in \overline{E}$.

Lemma 5.2 *If f_1 and f_2 are H-integrable on E, and if $0 \leq f_i(x) \leq h(x)$ almost everywhere in $\overline{E}$ for $i = 1, 2$ where h is also H-integrable on E, then the function $\max(f_1, f_2)$ is H-integrable on E.*

Proof.   For each subinterval I of $\overline{E}$, let $F_i(I)$ denote the integral of f_i on I for $i = 1, 2$ and let

$$F^*(I) = \max\left(F_1(I), F_2(I)\right).$$

Take any division $D = \{(I, x)\}$ of E, we have

$$0 \leq (D) \sum F^*(I) \leq (H) \int_E h.$$

Denote by A the supremum of all such sums $(D) \sum F^*(I)$. We shall show that A is the integral of the function $\max(f_1, f_2)$ on E. Given $\varepsilon > 0$, there is a gauge δ on E such that for any δ-fine division $D = \{(I, x)\}$ of E, we have, for $i = 1, 2$,

$$(D) \sum |f_i(x)\iota(I) - F_i(I)| < \varepsilon.$$

Now for each subinterval J of $\overline{E}$, and for $i = 1, 2$, let

$$S_i(J) = \sup\left((D) \sum |f_i(x)\iota(I) - F_i(I)|\right),$$

where the supremum is over all δ-fine divisions $D = \{(I, x)\}$ of J. Note that for $i = 1, 2$, we have $S_i(E) \leq \varepsilon$ and if $J = J_1 \cup J_2$ where J_1 and J_2 are non-overlapping subintervals of E, then

$$S_i(J_1) + S_i(J_2) \leq S_i(J).$$

In contrast,

$$F^*(J) \leq F^*(J_1) + F^*(J_2).$$

Furthermore, for any δ-fine division $D = \{(I, x)\}$ of E, we have, for $i = 1, 2$,

$$f_i(x)\iota(I) \leq F^*(I) + S_1(I) + S_2(I).$$

Consequently, writing $f = \max(f_1, f_2)$ we have

$$f(x)\iota(I) \leq F^*(I) + S_1(I) + S_2(I).$$

Similarly, we also have

$$F^*(I) - S_1(I) - S_2(I) \leq f(x)\iota(I).$$

Combining the above two inequalities and applying the finite superadditivity of S_1 and S_2, we obtain

$$\left| (D) \sum [f(x)\iota(I) - F^*(I)] \right| \leq (D) \sum [S_1(I) + S_2(I)]$$
$$\leq S_1(E) + S_2(E)$$
$$\leq 2\varepsilon.$$

Finally, fix a division $D_0 = \{(J, y)\}$ such that its corresponding sum

$$(D_0) \sum F^*(J) > A - \varepsilon.$$

Modify δ in such a way that if $D = \{(I, x)\}$ is δ-fine, then it is a refinement of D_0, that is, every interval I of D is included in some interval J of D_0. For the modified gauge δ, we note that since F^* is finitely subadditive, for any δ-fine division $D = \{(I, x)\}$ of E, we have

$$0 \leq A - (D) \sum F^*(I)$$
$$\leq A - (D_0) \sum F^*(J)$$
$$< \varepsilon.$$

Applying the above inequalities we obtain

$$\left| (D) \sum f(x)\iota(I) - A \right| < 3\varepsilon.$$

Hence, we have proved that the function $\max(f_1, f_2)$ is H-integrable on E.   □

Lemma 5.3 *Let $f_1, f_2, \ldots, f_n$ be H-integrable functions on E. Suppose that $g(x) \leq f_i(x) \leq h(x)$ almost everywhere in $\overline{E}$ for $i = 1, 2, \ldots, n$ where g and h are also H-integrable on E. Then $\max(f_1, f_2, \ldots, f_n)$ and $\min(f_1, f_2, \ldots, f_n)$ are both H-integrable on E.*

Proof. We first prove that the functions $\max(f_1, f_2)$ and $\min(f_1, f_2)$ are both H-integrable on E. For $i = 1, 2$, since $g(x) \le f_i(x) \le h(x)$ almost everywhere in $\overline{E}$, we have

$$0 \le f_i(x) - g(x) \le h(x) - g(x)$$

for almost all x in $\overline{E}$ where $f_i - g$, $i = 1, 2$, and $h - g$ are H-integrable on E. By Lemma 5.2, the function $\max(f_1 - g, f_2 - g)$ is H-integrable on E. Since

$$\max(f_1 - g, f_2 - g) = \max(f_1, f_2) - g$$

where g is H-integrable on E, we infer that $\max(f_1, f_2)$ is H-integrable on E. On the other hand, note that

$$\min(f_1, f_2) = -\max(-f_1, -f_2)$$

and $-h(x) \le -f_i(x) \le -g(x)$ almost everywhere in $\overline{E}$ for $i = 1, 2$. Since the functions $-f_1$, $-f_2$, $-h$ and $-g$ are H-integrable on E, it follows that $\max(-f_1, -f_2)$ is also H-integrable on E and thus so is $\min(f_1, f_2)$. Finally, note that for each $i = 1, 2, \ldots, n - 2$,

$$\max(f_i, f_{i+1}, f_{i+2}) = \max(\max(f_i, f_{i+1}), f_{i+2})$$

and

$$\min(f_i, f_{i+1}, f_{i+2}) = \min(\min(f_i, f_{i+1}), f_{i+2}).$$

The H-integrability of

$$\max(f_1, f_2, \ldots, f_n) \quad \text{and} \quad \min(f_1, f_2, \ldots, f_n)$$

on E then follows by induction. $\qquad\square$

We shall state and prove the dominated convergence theorem. Note that given a sequence $\{a_n\}$ of real numbers, the lower limit, known as *limit inferior*, of the sequence is defined by

$$\liminf_{n \to \infty} a_n = \lim_{i \to \infty} \left[\inf_{n \ge i} a_n \right].$$

Similarly, the upper limit, known as *limit superior*, of the sequence $\{a_n\}$ is defined by

$$\limsup_{n \to \infty} a_n = \lim_{i \to \infty} \left[\sup_{n \ge i} a_n \right].$$

It is a well-known fact that $a_n \to a$ as $n \to \infty$ if and only if

$$\lim_{i \to \infty} \left[\inf_{n \geq i} a_n \right] = a = \lim_{i \to \infty} \left[\sup_{n \geq i} a_n \right].$$

These concepts will be used in the proof of the theorem which follows similar arguments as in [21].

Theorem 5.3 (Dominated Convergence Theorem). *Suppose that the following conditions are satisfied.*

(i) *A function f is such that $f_n(x) \to f(x)$ almost everywhere in $\overline{E}$ as $n \to \infty$ where each f_n is H-integrable on E;*

(ii) *For all n, $g(x) \leq f_n(x) \leq h(x)$ almost everywhere in $\overline{E}$ where g and h are H-integrable on E.*

Then f is H-integrable on E and

$$(H) \int_E f_n \to (H) \int_E f \quad as \ n \to \infty.$$

Proof. By Lemma 5.3, for any positive integers i and j the function $\min\{f_n : i \leq n \leq j\}$ is H-integrable on E. For each fixed i, we then let

$$f_j^* = \min\{f_n : i \leq n \leq j\}$$

for $j = i, i+1, i+2, \ldots$. Note that the sequence of functions $-f_i^*, -f_{i+1}^*, \ldots$ is monotone increasing and the integrals of the functions are bounded above by $-(H) \int_E g$. By the monotone convergence theorem (Theorem 1.2), the limit function $\inf\{f_n : n \geq i\}$ is H-integrable on E. Similarly, we can show that $\sup\{f_n : n \geq i\}$ is H-integrable on E. As a result, we have

$$(H) \int_E \left(\inf_{n \geq i} f_n \right) \leq \inf_{n \geq i} \left((H) \int_E f_n \right) \tag{5.3}$$

$$\leq \sup_{n \geq i} \left((H) \int_E f_n \right) \leq (H) \int_E \left(\sup_{n \geq i} f_n \right).$$

Next, for $i = 1, 2, \ldots$, we let f_i^{**} denote the function $\inf\{f_n : n \geq i\}$. Clearly the sequence $\{f_i^{**}\}_{i=1}^{\infty}$ is monotone increasing and the integrals of f_i^{**} are bounded above by $(H)\int_E h$. Furthermore, $f_i^{**}(x) \to f(x)$ as $i \to \infty$ almost everywhere in $\overline{E}$. So by applying the monotone convergence theorem again, this time to the sequence $\{f_i^{**}\}$, we prove that f is H-integrable on E and

$$\lim_{i \to \infty}\left[(H)\int_E \left(\inf_{n \geq i} f_n\right)\right] = (H)\int_E f.$$

By the same token, we can prove that

$$\lim_{i \to \infty}\left[(H)\int_E \left(\sup_{n \geq i} f_n\right)\right] = (H)\int_E f.$$

Consequently, by taking limit as $i \to \infty$ in (5.3), we obtain

$$(H)\int_E f \leq \lim_{i \to \infty}\left[\inf_{n \geq i}\left((H)\int_E f_n\right)\right]$$
$$\leq \lim_{i \to \infty}\left[\sup_{n \geq i}\left((H)\int_E f_n\right)\right] \leq (H)\int_E f,$$

that is,

$$\lim_{i \to \infty}\left[\inf_{n \geq i}\left((H)\int_E f_n\right)\right] = (H)\int_E f = \lim_{i \to \infty}\left[\sup_{n \geq i}\left((H)\int_E f_n\right)\right].$$

It follows that

$$\lim_{n \to \infty}(H)\int_E f_n = (H)\int_E f$$

and the proof is complete. $\qquad\qquad\square$

The above proof of the dominated convergence theorem has in fact also proved the following result which is known as Fatou's lemma.

Corollary 5.2 (Fatou's Lemma). *Let f_n, $n = 1, 2, \ldots$, be non-negative and H-integrable on E with $f_n(x) \to f(x)$ almost everywhere in $\overline{E}$ as $n \to \infty$. If the sequence of the integrals of $f_1, f_2, \ldots$ is bounded, then f is H-integrable on E and*

$$(H) \int_E f \le \liminf_{n \to \infty} \left((H) \int_E f_n \right).$$

The dominated convergence theorem can be applied to prove the following result.

Proposition 5.1 *Let W be a measurable subset of $\overline{E}$. Then the function χ_W is H-integrable on E to the value $\iota(W)$.*

Proof. By Condition $(*)$, for each positive integer n, there exists a closed set V_n such that $V_n \subseteq W$ and $\iota(W \setminus V_n) < \dfrac{1}{n}$. We then define $f_n = \chi_{V_n}$ for each $n = 1, 2, \ldots$. Since V_n is closed, by Proposition 1.7, each f_n is H-integrable on E to $\iota(V_n)$. Clearly, $f_n(x) \to \chi_W(x)$ as $n \to \infty$ almost everywhere in $\overline{E}$. Now define $g(x) = 0$ and $h(x) = 1$ for each $x \in \overline{E}$. Then g and h are H-integrable on E and

$$g(x) \le f_n(x) \le h(x)$$

for each $x \in \overline{E}$. By the dominated convergence theorem (Theorem 5.3), χ_W is H-integrable on E and

$$(H) \int_E \chi_W = \lim_{n \to \infty} (H) \int_E f_n$$

where

$$\lim_{n \to \infty} (H) \int_E f_n = \lim_{n \to \infty} \iota(V_n)$$
$$= \iota(W) - \lim_{n \to \infty} \iota(W \setminus V_n).$$

Since $\iota(W \setminus V_n) < \dfrac{1}{n}$ which tends to 0 as $n \to \infty$, it follows that $(H) \int_E \chi_W = \iota(W)$. This completes the proof. $\qquad\square$

The following result will play a crucial role in the proof of the Harnack convergence theorem.

Theorem 5.4 (Mean Convergence Theorem). *Let W be a measurable subset of $\overline{E}$. Suppose that $f_n(x) \to f(x)$ almost everywhere in $\overline{E}$ as $n \to \infty$ where f_n, $n = 1, 2, \ldots$, is H-integrable on W and satisfies the condition that*

$$(H) \int_W |f_n - f_m| \to 0 \quad as \ n, m \to \infty.$$

Then f is H-integrable on W and

$$(H) \int_W f_n \to (H) \int_W f \quad as \ n \to \infty.$$

Proof. Since $(H) \int_W |f_n - f_m| \to 0$ as $n, m \to \infty$, we can choose positive integers $n(1) < n(2) < \cdots$ such that

$$(H) \int_W \left| f_{n(i)} - f_{n(i+1)} \right| < 2^{-i}$$

for $i = 1, 2, \ldots$. Define a function g on $\overline{E}$ given by

$$g(x) = \sum_{i=1}^{\infty} \left| f_{n(i)}(x) - f_{n(i+1)}(x) \right|$$

if $x \in W$ and 0 otherwise. Then by the monotone convergence theorem (Theorem 1.2), the function g exists almost everywhere in $\overline{E}$ and is H-integrable on E. Therefore the sequence $\{f_{n(i)}\chi_W\}_{i=1}^{\infty}$ of functions is dominated on the right by $f_{n(1)}\chi_W + g$ and on the left by $f_{n(1)}\chi_W - g$. By the dominated convergence theorem (Theorem 5.3), the function $f\chi_W$ is H-integrable on E. On the other hand, for each fixed n, applying Fatou's lemma (Corollary 5.2) on the sequence $\{|f_m - f_n|\}$ in m yields

$$(H) \int_W |f_n - f| \leq \liminf_{m \to \infty} \left[(H) \int_W |f_m - f_n| \right]$$

where the term on the right-hand side of the inequality tends to 0 as $n \to \infty$. It follows that $(H) \int_W |f_n - f| \to 0$ as $n \to \infty$ and thus

$$(H) \int_W f_n \to (H) \int_W f \quad \text{as } n \to \infty$$

as desired.   □

Note that the mean convergence theorem (Theorem 5.4) involves the convergence of $\{f_n\}$ whereas the generalised mean convergence theorem (Theorem 3.12) involves the convergence of $\{F_n\}$.

We shall next prove the Harnack convergence theorem by means of the category argument. Let the functions $f_1, f_2, \ldots$ be H-integrable on E and let f be a function such that $f_n(x) \to f(x)$ almost everywhere in $\overline{E}$ as $n \to \infty$. We say that a point x in $\overline{E}$ is *regular* if there is an interval $I_x \subseteq \overline{E}$, which is dependent on x, with $x \in I_x^o$, where I_x^o denotes the interior of I_x, such that f is H-integrable on I_x and

$$(H) \int_{I_x} f_n \to (H) \int_{I_x} f \quad \text{as } n \to \infty.$$

Clearly, if I_x is an interval as given in the above definition of a regular point, then every point in I_x^o is regular. It follows that the set of all regular points in $\overline{E}$ is an open set and hence the set of all points in $\overline{E}$ which are not regular is a closed set. Furthermore, if all points in $\overline{E}$ are regular, then

$$\bigcup_{x \in \overline{E}} I_x = \overline{E}.$$

Since $\overline{E}$ is compact, there exists $x_1, x_2, \ldots, x_n \in \overline{E}$ such that $\bigcup_{i=1}^{n} I_{x_i} = \overline{E}$. By definition, the function f is H-integrable on each of the finitely many intervals I_{x_i}. Consequently, it is H-integrable on E and

$$(H) \int_E f_n \to (H) \int_E f \quad \text{as } n \to \infty.$$

We are now ready to prove the Harnack convergence theorem which is formulated as follows.

Theorem 5.5 (Harnack Convergence Theorem). *Let the functions $f_1, f_2, \ldots$ be H-integrable on E and $X_1, X_2, \ldots$ be closed sets whose union is $\overline{E}$. Suppose that the following conditions are satisfied.*

(i) *A function f is such that $f_n(x) \to f(x)$ almost everywhere in $\overline{E}$ as $n \to \infty$;*

(ii) *For each $i = 1, 2, \ldots$, for every closed set $Y \subseteq X_i$ and for every interval I such that $Y \subseteq I$, the sequence $\{f_n\}$ satisfies the uniform Harnack condition on Y with respect to I;*

(iii) *For each $i = 1, 2, \ldots$ and for every closed set $Y \subseteq X_i$, the sequence $\{f_n\}$ satisfies the uniform (AC)-condition on Y;*

(iv) *There exists a gauge δ on E such that for any δ-fine interval-point pair (I, x), the sequence $\{F_n(I)\}$ converges as $n \to \infty$.*

Then f is H-integrable on E and

$$(H) \int_E f_n \to (H) \int_E f \quad \text{as } n \to \infty.$$

Proof.   Let S be the set of all points in $\overline{E}$ that are not regular. Then S is a closed subset of $\overline{E}$. It suffices to prove that S is empty. We shall prove by contradiction and so we suppose $S \neq \emptyset$. Then, since

$$S = \bigcup_{i=1}^{\infty} (S \cap X_i),$$

where for each i, the set $S \cap X_i$ is closed, by the Baire category theorem (Theorem 5.2), there exists a positive integer m such that $(S \cap X_m)^o \neq \emptyset$. Let $x_0 \in (S \cap X_m)^o$. Since $(S \cap X_m)^o$ is open, there exists an open ball B centred at x_0 such that

$$B \subseteq (S \cap X_m)^o \cap B(x_0, \delta(x_0)),$$

where δ is the gauge given in hypothesis (iv). Choose an open ball B_0, also centred at x_0, whose radius is strictly less than that of B. Then $\overline{B_0} \subseteq B$ and we let $I_0 = \overline{B_0}$. By the definition of $\mathcal{I}_1^*$, the set I_0 is a generalised interval. Also note that (I_0, x_0) is δ-fine and

$$I_0 \cap S \subseteq I_0 \subseteq (S \cap X_m)^o \subseteq X_m.$$

Now let Y denote the closed set $I_0 \cap S$. We shall prove that f is H-integrable on I_0 which leads to a contradiction. Let $\varepsilon > 0$ be given. By hypothesis (iii), the sequence $\{f_n\}$ satisfies the uniform (AC)-condition on Y, and so condition (5.2) in Definition 5.4 involving $\eta > 0$ holds. Then by Egoroff's theorem (Theorem 3.7), there is a set $W \subseteq Y$ such that f_n converges to f uniformly on W with $\iota(Y \setminus W) < \eta$. Thus, there exists a positive integer M such that whenever $n, m \geq M$, we have $|f_n(x) - f_m(x)| < \varepsilon$ for all $x \in W$. Consequently, for any $n, m \geq M$, we have

$$(H) \int_Y |f_n - f_m| \leq (H) \int_W |f_n - f_m| + (H) \int_{Y \setminus W} |f_n|$$
$$+ (H) \int_{Y \setminus W} |f_m|$$
$$< \varepsilon \times \iota(Y) + 2\varepsilon.$$

It follows from the mean convergence theorem (Theorem 5.4) that f is H-integrable on Y and

$$(H) \int_Y f_n \to (H) \int_Y f \quad \text{as } n \to \infty.$$

Next, if $I_0 \setminus Y = \emptyset$, we are done. So suppose $I_0 \setminus Y \neq \emptyset$ and let J be an interval which is a subset of $I_0 \setminus Y$. It is easy to see that J does not contain any point which is not regular. Indeed, if J contains a point $x \in S$, then since $J \subseteq I_0$, it follows that $x \in I_0 \cap S = Y$ which is a contradiction because x belongs to J and thus is in $I_0 \setminus Y$. So since every point in J is regular, the function f is H-integrable on J and

$$(H) \int_J f_n \to (H) \int_J f \quad \text{as } n \to \infty.$$

Since an elementary set is a finite union of disjoint intervals, we can conclude that f is H-integrable on every elementary subset E^* of $I_0 \setminus Y$ and

$$(H) \int_{E^*} f_n \to (H) \int_{E^*} f \quad \text{as } n \to \infty.$$

Now, since $Y \subseteq I_0 \subseteq X_m$, by hypothesis (ii), the sequence $\{f_n\}$ satisfies the uniform Harnack condition on Y with respect to I_0. So there exist real numbers A_n, $n = 1, 2, \ldots$, such that for every $\varepsilon > 0$ given, there exists a gauge δ on I_0, independent of n, such that for any nonabsolute subset E_1 of $I_0 \setminus Y$ involving δ, we have

$$\left| (H) \int_{E_1} f_n - A_n \right| < \varepsilon$$

for all n. By Theorem 5.1, for each $n = 1, 2, \ldots$, we have

$$(H) \int_{I_0} f_n = (H) \int_Y f_n + A_n.$$

Hence we obtain

$$A_n = (H) \int_{I_0} f_n - (H) \int_Y f_n$$

$$= F_n(I_0) - (H) \int_Y f_n$$

where $\lim_{n \to \infty} (H) \int_Y f_n = (H) \int_Y f$. Since (I_0, x_0) is δ-fine, by hypothesis (iv), the sequence $\{F_n(I_0)\}$ is convergent in n and thus, so is $\{A_n\}$. Let A be the limit of $\{A_n\}$ and let N be a

positive integer such that
$$|A_n - A| < \varepsilon$$
whenever $n \geq N$. We may assume that N is large enough so that for any nonabsolute subset E_1 of $I_0 \setminus Y$ involving δ, we also have
$$\left| (H) \int_{E_1} f_N - (H) \int_{E_1} f \right| < \varepsilon.$$
Hence we obtain
$$\begin{aligned}
\left| (H) \int_{E_1} f - A \right| &\leq \left| (H) \int_{E_1} f_N - (H) \int_{E_1} f \right| \\
&\quad + \left| (H) \int_{E_1} f_N - A_N \right| \\
&\quad + |A_N - A| \\
&< 3\varepsilon
\end{aligned}$$
and so f satisfies the Harnack condition on Y with respect to I_0. By Theorem 5.1, the function f is H-integrable on I_0 and
$$(H) \int_{I_0} f_n \to (H) \int_{I_0} f \quad \text{as } n \to \infty.$$
Therefore, all points in the interior of I_0, including x_0, are regular. However, this is a contradiction as $x_0 \in (S \cap X_m)^o$ which means x_0 is not a reglar point. It follows that $S = \emptyset$ and hence every point in E is regular. In other words, the function f is H-integrable on E and
$$(H) \int_E f_n \to (H) \int_E f \quad \text{as } n \to \infty.$$
This completes the proof.

$\square$

We have therefore recovered, for the H-integral, the proof by means of category argument in proving convergence theorems. In classical integration theory, both Cauchy extension and Harnack extension have to be applied in the category proof. In the proof of Theorem 5.5, we see that applying just the Cauchy–Harnack extension alone is sufficient.

5.4   An Improved Version of the Controlled Convergence Theorem

In this section, we shall show that an improved version of the controlled convergence theorem for the H-integral follows as a consequence of the Harnack convergence theorem. We first recall a definition related to generalised absolute continuity.

The following definition is actually the metric space analogue of Definition 3.5. Note that the condition "F is $ABRS$ on Y" in the latter is dropped here as it is superfluous when $\mathcal{T}$ is a metric topology. We state the definition here for the convenience of the reader.

Definition 5.5   Let $Y \subseteq \overline{E}$ be a measurable set.   An elementary-set function F is said to be $AC^{\Delta}(Y)$ if for every $\varepsilon > 0$ there exist a gauge δ on E and $\eta > 0$ such that for any two δ-fine partial divisions $D_1 = \{(I, x)\}$ and $D_2 = \{(I, y)\}$ of E with x, $y \in Y$ such that D_2 is a refinement of D_1 satisfying the condition that $(D_1 \setminus D_2) \sum \iota(I) < \eta$, we have

$$\left| (D_1 \setminus D_2) \sum F(I) \right| < \varepsilon.$$

Here D_2 may be void. If $\overline{E}$ is the union of closed sets X_i, $i = 1, 2, \ldots$ such that F is $AC^{\Delta}(X_i)$ for each i, then F is said to be ACG^{Δ} on E.

It is clear from the definition that if F is $AC^{\Delta}(Y)$, then F is in fact $AC^{\Delta}(W)$ for every measurable subset W of Y.

Let E_1 be the union of the component intervals of $D_1 = \{(I, x)\}$ and E_2 be the union of the component intervals of $D_2 = \{(J, y)\}$. Since D_2 is a refinement of D_1, we have $E_2 \subseteq E_1$. If F is finitely additive over intervals, then

$$(D_1 \setminus D_2) \sum F(I) = (D_1) \sum F(I) - (D_2) \sum F(J)$$
$$= F(E_1) - F(E_2).$$

Likewise, the following definition is the metric space analogue of Definition 3.10. Again note that the $ABRS$ condition is omitted here because when the topology $\mathcal{T}$ is metrizable, the condition is superfluous.

Definition 5.6   Let $Y \subseteq \overline{E}$ be a measurable set. We say that a sequence $\{F_n\}$ of elementary-set functions is $UAC^\Delta(Y)$ if F_n is $AC^\Delta(Y)$ uniformly in n, that is the gauge δ in Definition 5.5 is independent of n. If $\overline{E}$ is the union of closed sets X_i, $i = 1, 2, \ldots$ such that $\{F_n\}$ is $UAC^\Delta(X_i)$ for each i, then $\{F_n\}$ is said to be $UACG^\Delta$ on E.

Recall that a function is absolutely H-integrable on a measurable set Y if and only if it is Lebesgue integrable on Y. Then applying Theorem 3.4 yields the following result.

Lemma 5.4   *Let f be an H-integrable function on E with primitive F and let $Y \subseteq \overline{E}$ be a closed set. If F is $AC^\Delta(Y)$, then for every measurable subset W of Y, the function f_W is absolutely H-integrable on E. Furthermore, the primitive F_Y of f_Y is $AC^\Delta(Y)$.*

By virtue of Lemma 3.2 and Lemma 5.4, we obtain the following result.

Lemma 5.5   *Let $Y \subseteq \overline{E}$ be a closed set and let $f, f_1, f_2, \ldots$ be real-valued functions on $\overline{E}$. Suppose that the following conditions are satisfied.*

(i) *$f_{n,Y}(x) \to f_Y(x)$ almost everywhere in $\overline{E}$ as $n \to \infty$ where each $f_{n,Y}$ is H-integrable on E with primitive $F_{n,Y}$;*
(ii) *The sequence $\{F_{n,Y}\}$ is $UAC^\Delta(Y)$.*

Then the function f_Y is H-integrable on E with integral value $F_Y(E)$ satisfying

$$\lim_{n \to \infty} F_{n,Y}(E) = F_Y(E)$$

and

$$(H) \int_Y |f_n - f| \to 0 \quad as \; n \to \infty.$$

What follows is the metric space analogue of Lemma 3.3.

Lemma 5.6 *Let f_n, $n = 1, 2, \ldots$, be H-integrable functions on E with primitive F_n and let $Y \subseteq \overline{E}$ be a closed set. If $\{F_n\}$ is $UAC^\Delta(Y)$, then for every $\varepsilon > 0$ there exists a gauge δ on E, independent of n, such that for every δ-fine partial division $D = \{(I, x)\}$ with $x \in Y$, we have, for all n,*

$$(D) \sum |F_{n,Y}(I) - F_n(I)| < \varepsilon.$$

We can now prove the following improved version of the controlled convergence theorem for the H-integral.

Theorem 5.6 (Controlled Convergence Theorem). *Let $\{X_i\}$ be a sequence of closed sets whose union is $\overline{E}$. Suppose that the following conditions are satisfied.*

(i) *$f_n(x) \to f(x)$ almost everywhere in $\overline{E}$ as $n \to \infty$ where for $n = 1, 2, \ldots$, the function f_n is H-integrable on E with primitive F_n;*

(ii) *The sequence $\{F_n\}$ is $UACG^\Delta$ on E;*

(iii) *There exists a gauge δ on E such that for any δ-fine interval-point pair (I, x), the sequence $\{F_n(I)\}$ converges as $n \to \infty$.*

Then f is H-integrable on E and

$$\lim_{n \to \infty} (H) \int_E f_n = (H) \int_E f.$$

Proof. It suffices to prove that conditions (ii) and (iii) in the Harnack convergence theorem (Theorem 5.5) are satisfied. First note that since the sequence $\{F_n\}$ is $UACG^\Delta$ on E, for each

i, it is $UAC^\Delta(X_i)$. Now let i be fixed, then let Y be a closed subset of X_i and J be an interval such that $Y \subseteq J$. By virtue of Lemma 5.4, we see that each f_n is absolutely H-integrable on every measurable subset W of Y. For each n, since f_n is H-integrable on E, it is H-integrable on J and every nonabsolute subset E_1 of $J \setminus Y$. To prove that $\{f_n\}$ satisfies the uniform Harnack condition on Y with respect to J, we let $\varepsilon > 0$ be given and apply Lemma 5.6 to choose a gauge δ_i on J, independent of n, such that for any δ_i-fine partial division $D = \{(I, x)\}$ with $x \in Y$, we have

$$(D) \sum |F_{n,Y}(I) - F_n(I)| < \varepsilon$$

where $F_{n,Y}$ denotes the primitive of $f_{n,Y}$. Modify the gauge δ_i if necessary so that $B(x, \delta_i(x)) \cap Y = \emptyset$ if $x \notin Y$. Then let E_1 be a nonabsolute subset of $J \setminus Y$ involving δ_i, and let $U = J \setminus E_1$. Note that U is a δ-fine cover of Y by definition. Let $D_0 = \{(K, \xi)\}$ be the δ_i-fine partial division of J such that U is the finite union of the component intervals K of D_0. Clearly, for each $(K, \xi) \in D_0$, we must have $\xi \in Y$ which implies that

$$(D_0) \sum |F_{n,Y}(K) - F_n(K)| < \varepsilon.$$

Note that

$$(D_0) \sum F_{n,Y}(K) = (D_0) \sum \left[(H) \int_K f_{n,Y} \right]$$

$$= (H) \int_E f_{n,Y} \chi_U$$

$$= (H) \int_Y f_n$$

while

$$(D_0) \sum F_n(K) = (D_0) \sum \left[(H) \int_K f_n \right]$$

$$= (H) \int_U f_n.$$

Hence we have, for all n,

$$\left| (H) \int_Y f_n - (H) \int_U f_n \right| \tag{5.4}$$
$$= \left| (D_0) \sum F_{n,Y}(K) - (D_0) \sum F_n(K) \right|$$
$$\leq (D_0) \sum |F_{n,Y}(K) - F_n(K)|$$
$$< \varepsilon.$$

Next, since each f_n is H-integrable on E and on Y, it is H-integrable on $J \setminus Y$. So for each n, we let

$$A_n = (H) \int_{J \setminus Y} f_n.$$

Note that

$$(H) \int_J f_n = (H) \int_{E_1} f_n + (H) \int_U f_n$$
$$= (H) \int_{J \setminus Y} f_n + (H) \int_Y f_n$$
$$= A_n + (H) \int_Y f_n$$

from which we infer that

$$\left| (H) \int_{E_1} f_n - A_n \right| = \left| (H) \int_Y f_n - (H) \int_U f_n \right|.$$

It then follows from (5.4) that for all n, we have

$$\left| (H) \int_{E_1} f_n - A_n \right| < \varepsilon.$$

Hence $\{f_n\}$ satisfies the uniform Harnack condition on Y with respect to J. Finally, since each $f_{n,Y}$ is absolutely H-integrable on E and the sequence $\{F_{n,Y}\}$ is $UAC^\Delta(Y)$, by Lemma 5.5, the function f_Y is also absolutely H-integrable on E and

$$(H) \int_Y |f_n - f| \to 0 \quad \text{as } n \to \infty.$$

Since f_Y is absolutely H-integrable on E, it satisfies the (AC)-condition on Y. Thus, there exists $\eta > 0$ such that whenever $W \subseteq Y$ and $\iota(W) < \eta$, we have

$$(H) \int_W |f| < \varepsilon.$$

Consequently, for all sufficiently large n, we have

$$(H) \int_W |f_n| \leq (H) \int_W |f_n - f| + (H) \int_W |f|$$
$$\leq (H) \int_Y |f_n - f| + (H) \int_W |f|$$
$$< 2\varepsilon.$$

Therefore, the sequence $\{f_n\}$ satisfies the uniform (AC)-condition on Y. This completes the proof. $\qquad\square$

Let us now compare the two versions of controlled convergence theorems we have proved in this book, namely Theorem 3.15 and the above theorem. Note that hypotheses (i) and (ii) of Theorem 3.15 are required in the above theorem. However, while hypothesis (iii) of Theorem 3.15 requires the existence of an elementary-set function F on E, which is finitely additive over intervals, and a gauge G on E such that for any G-fine interval-point pair (I, x), we have $F_n(I) \to F(I)$ as $n \to \infty$, in the above theorem we only require the existence of a gauge δ such that for each δ-fine interval-point pair (I, x), the sequence $\{F_n(I)\}$ is convergent in n without requiring the existence of an elementary-set function F on E. Hence the above theorem is an improved version of Theorem 3.15.

We remark that in the controlled convergence theorem of the HK-integral in the n-dimensional Euclidean space, the convergence of the sequence $\{F_n\}$ of primitives, where each F_n is an elementary-set function, can be made redundant. This is possible because if $\{F_n\}$ is $UACG^{**}$, the Euclidean space analogue of the $UACG^\triangle$ condition, then by the well-known Ascoli–Arzela theorem, the sequence $\{F_n\}$, where each F_n is a point function, has a uniformly convergent subsequence. However, in our setting, each primitive F_n is an elementary-set function and there is no known method of converting F_n from an elementary-set function to a point function in a mutually convertible manner as in the case of the Euclidean space (see Section 4.3 or [21, page 138]). As a result, in both versions of controlled convergence theorem in this book, we still require a hypothesis involving the convergence of $\{F_n\}$.

The final result we shall prove is one that relates the uniform Harnack condition to H-equiintegrability. We first prove the following lemma.

Lemma 5.7 *Let $\{f_n\}$ be a sequence of H-integrable functions on E and $\{X_i\}$ be a monotone increasing sequence of closed sets whose union is $\overline{E}$. Suppose that the following conditions are satisfied.*

(i) *For each x in $\overline{E}$, the sequence $\{f_n(x)\}$ converges as $n \to \infty$;*

(ii) *For each $i = 1, 2, \ldots$, and for every closed set $Y \subseteq X_i$, the sequence $\{f_n\}$ satisfies the uniform (AC)-condition on Y.*

Then for each $i = 1, 2, \ldots$, the sequence $\{f_{n,X_i}\}$, where $f_{n,X_i} := f_n \chi_{X_i}$, is H-equiintegrable on E.

Proof. Let $\varepsilon > 0$ be given. We shall prove that for each i, there exists a gauge δ_i on E, independent of n, such that for any δ_i-fine partial division $D = \{(I, x)\}$ of E with associated points $x \in X_i$, the inequality

$$(D) \sum \left| f_n(x) \iota(I) - (H) \int_I f_{n,X_i} \right| < \varepsilon$$

holds for all n. First, since for each positive integer i and for every closed set $Y \subseteq X_i$, the sequence $\{f_n\}$ satisfies the uniform (AC)-condition on Y, for each X_i and for any positive integer j, there exists $\eta_{i,j} > 0$ such that for all measurable subsets W of X_i satisfying the condition that $\iota(W) < \eta_{i,j}$, we have

$$(H) \int_W |f_n| < \frac{\varepsilon}{14 \times 2^{i+j}} \tag{5.5}$$

which holds for all n. Next, since the sequence $\{f_n\}$ of functions is pointwise convergent on $\overline{E}$, by Egoroff's theorem (Theorem 3.7), for each positive integer j, there exists a closed subset Y_j of $\overline{E}$ such that $f_n \to f$ uniformly on Y_j and

$$\iota \left(\overline{E} \setminus \bigcup_{j=1}^{\infty} Y_j \right) = 0.$$

Now let a positive integer i_0 be fixed. For each $j = 1, 2, \ldots$, define $W_j = X_{i_0} \cap Y_j$, which is obviously a closed set, and choose a positive integer N_j such that

$$|f_m(x) - f_n(x)| < \frac{\varepsilon}{14 \times \iota(E) \times 2^j} \tag{5.6}$$

for each $x \in W_j$ and $m, n > N_j$. Also find an open set U_j such that $W_j \subseteq U_j$ and $\iota(U_j \setminus W_j) < \eta_{i,j}$. Furthermore, since each f_n is H-integrable on X_{i_0}, by Henstock's lemma, there exists a gauge δ_n on E such that for all δ_n-fine partial divisions of E with associated points $x \in W_j$, we have

$$(D) \sum \left| f_n(x)\iota(I) - (H) \int_I f_{n,X_{i_0}} \right| < \frac{\varepsilon}{14 \times 2^j}. \tag{5.7}$$

We may assume that for each n, the gauge δ_{n+1} is finer than δ_n. Now define a gauge δ on E such that for each j, if $x \in W_j$, then

$$B(x, \delta(x)) \subseteq U_j \cap B(x, \delta_{N_j}(x)).$$

This ensures that the gauge δ is finer than δ_{N_j}. Let $D = \{(I_k, x_k) : k = 1, 2, \ldots, m\}$ be a δ-fine partial division of E with associated points $x \in X_{i_0}$ and let a positive integer n be fixed. Note that

$$(D) \sum \left| f_n(x) \iota(I) - (H) \int_I f_{n,X_{i_0}} \right|$$

$$\leq (D) \sum_{N_j \geq n} \left| f_n(x) \iota(I) - (H) \int_I f_{n,X_{i_0}} \right|$$

$$+ (D) \sum_{N_j < n} \left| f_n(x) \iota(I) - (H) \int_I f_{n,X_{i_0}} \right|$$

where $\sum\limits_{N_j \geq n}$ sums over all indices j such that $N_j \geq n$ and $\sum\limits_{N_j < n}$ sums over all indices j such that $N_j < n$. Note that if $N_j \geq n$, then δ_{N_j} is finer than δ_n. Hence, by (5.7), we have

$$(D) \sum_{N_j \geq n} \left| f_n(x) \iota(I) - (H) \int_I f_{n,X_{i_0}} \right| < \sum_{j=1}^{\infty} \frac{\varepsilon}{14 \times 2^j}$$

$$< \frac{\varepsilon}{2}.$$

It remains to prove that

$$(D) \sum_{N_j < n} \left| f_n(x) \iota(I) - (H) \int_I f_{n,X_{i_0}} \right| < \frac{\varepsilon}{2}.$$

Let δ^* be a gauge on E which is finer than both δ and δ_n such that for each j, if $y \in W_j$, then $B(y, \delta^*(y)) \subseteq U_j$ and if $y \in U_j \setminus W_j$, then $B(y, \delta^*(y)) \subseteq U_j \setminus W_j$. For each $(I_k, x_k) \in D$ such that $x_k \in W_j$, we apply Cousin's lemma (Theorem 1.1) to find a δ^*-fine division $D_{(k)} = \{(J, y)\}$ of I_k. Let

$$D_{(k)}^+ = \{(J, y) \in D_{(k)} : y \in W_j\}$$

and

$$D_{(k)}^- = \{(J, y) \in D_{(k)} : y \notin W_j\},$$

and define

$$E_j = \bigcup_{x_k \in W_j} \bigcup_{(J,y) \in D_{(k)}^-} J.$$

Since each $I_k \subseteq B(x_k, \delta(x_k)) \subseteq U_j$ if $x_k \in W_j$, it follows that $E_j \subseteq U_j \setminus W_j$. For each $i = 1, 2, \ldots$, let $V_{i,j} = X_i \cap E_j$. Then $V_{i,j} \subseteq X_i$ and

$$\iota(V_{i,j}) \le \iota(U_j \setminus W_j) < \eta_{i,j}.$$

Thus, by inequality (5.5), for each i, we have

$$(H) \int_{V_{i,j}} |f_{n,X_{i_0}}| \le (H) \int_{V_{i,j}} |f_n|$$
$$< \frac{\varepsilon}{14 \times 2^{i+j}}$$

for all n which yields

$$(H) \int_{E_j} |f_{n,X_{i_0}}| \le \sum_{i=1}^{\infty} \left[(H) \int_{V_{i,j}} |f_{n,X_{i_0}}| \right]$$
$$< \sum_{i=1}^{\infty} \frac{\varepsilon}{14 \times 2^{i+j}}$$
$$= \frac{\varepsilon}{14 \times 2^j}.$$

Consequently, for all n, we obtain

$$\sum_{N_j < n} \sum_{x_k \in W_j} \sum_{(J,y) \in D^-_{(k)}} \left[(H) \int_J |f_{n,X_{i_0}}| \right] \tag{5.8}$$
$$\le \sum_{j=1}^{\infty} \left[(H) \int_{E_j} |f_{n,X_{i_0}}| \right]$$
$$= \frac{\varepsilon}{14}.$$

In particular, we have

$$\sum_{N_j < n} \sum_{x_k \in W_j} \sum_{(J,y) \in D^-_{(k)}} \left[(H) \int_J |f_{N_j,X_{i_0}}| \right] \le \frac{\varepsilon}{14}. \tag{5.9}$$

Note that by the triangle inequality,

$$\left| f_n(x_k)\iota(I_k) - (H)\int_{I_k} f_{n,X_{i_0}} \right|$$

$$\leq \left| f_n(x_k)\iota(I_k) - f_{N_j}(x_k)\iota(I_k) \right|$$

$$+ \left| f_{N_j}(x_k)\iota(I_k) - (H)\int_{I_k} f_{N_j,X_{i_0}} \right|$$

$$+ \left| (H)\int_{I_k} f_{n,X_{i_0}} - (H)\int_{I_k} f_{N_j,X_{i_0}} \right|$$

and hence

$$\sum_{N_j < n} \sum_{x_k \in W_j} \left| f_n(x_k)\iota(I_k) - (H)\int_{I_k} f_{n,X_{i_0}} \right|$$

$$\leq \sum_{N_j < n} \sum_{x_k \in W_j} \left| f_n(x_k) - f_{N_j}(x_k) \right| \iota(I_k)$$

$$+ \sum_{N_j < n} \sum_{x_k \in W_j} \left| f_{N_j}(x_k)\iota(I_k) - (H)\int_{I_k} f_{N_j,X_{i_0}} \right|$$

$$+ \sum_{N_j < n} \sum_{x_k \in W_j} \sum_{(J,y) \in D_{(k)}^+} \left| f_{N_j}(y)\iota(J) - (H)\int_J f_{N_j,X_{i_0}} \right|$$

$$+ \sum_{N_j < n} \sum_{x_k \in W_j} \sum_{(J,y) \in D_{(k)}^-} \left| (H)\int_J f_{N_j,X_{i_0}} \right|$$

$$+ \sum_{N_j < n} \sum_{x_k \in W_j} \sum_{(J,y) \in D_{(k)}^+} \left| f_n(y) - f_{N_j}(y) \right| \iota(J)$$

$$+ \sum_{N_j < n} \sum_{x_k \in W_j} \sum_{(J,y) \in D_{(k)}^+} \left| f_n(y)\iota(J) - (H)\int_J f_{n,X_{i_0}} \right|$$

$$+ \sum_{N_j < n} \sum_{x_k \in W_j} \sum_{(J,y) \in D_{(k)}^-} \left| (H)\int_J f_{n,X_{i_0}} \right|$$

where the first and the fifth terms on the right-hand side of the above inequality are each less than $\sum\limits_{j=1}^{\infty} \dfrac{\varepsilon}{14 \times \iota(E) \times 2^j} \times \iota(E)$ by inequality (5.6), while the second, third and sixth terms are each less than $\sum\limits_{j=1}^{\infty} \dfrac{\varepsilon}{14 \times 2^j}$ by inequality (5.7). Then in conjunction with inequalities (5.8) and (5.9), we obtain

$$\sum_{N_j < n}\sum_{x_k \in W_j} \left| f_n(x_k)\iota(I_k) - (H)\int_{I_k} f_{n,X_{i_0}} \right| < 7 \times \frac{\varepsilon}{14}$$

$$= \frac{\varepsilon}{2}$$

and therefore

$$(D) \sum \left| f_n(x)\iota(I) - (H)\int_I f_{n,X_{i_0}} \right| < \varepsilon \tag{5.10}$$

as desired. Finally, modify the gauge δ if necessary so that for each i, whenever $x \in \overline{E} \setminus X_{i_0}$ we have $B(x,\delta(x)) \cap X_{i_0} = \emptyset$, and let $D = \{(I,x)\}$ be a δ-fine division of E. We decompose D into $D_1 = \{(I,x) \in D : x \in X_{i_0}\}$ and $D_2 = \{(I,x) \in D : x \in \overline{E} \setminus X_{i_0}\}$. Obviously,

$$(D_1) \sum f_{n,X_{i_0}}(x)\iota(I) = (D_1) \sum f_n(x)\iota(I)$$

and

$$(D_2) \sum f_{n,X_{i_0}}(x)\iota(I) = 0.$$

Furthermore, since $I \cap X_{i_0} = \emptyset$ for each $(I,x) \in D_2$, we have

$$(D_2) \sum \left[(H)\int_I f_{n,X_{i_0}} \right] = 0$$

and hence

$$(D) \sum \left| f_{n,X_{i_0}}(x)\iota(I) - (H)\int_I f_{n,X_{i_0}} \right|$$

$$= (D_1) \sum \left| f_n(x)\iota(I) - (H)\int_I f_{n,X_{i_0}} \right|.$$

Consequently, using inequality (5.10) we obtain

$$\left| (D) \sum f_{n,X_{i_0}}(x)\iota(I) - (H) \int_{X_{i_0}} f_n \right|$$

$$= \left| (D) \sum f_{n,X_{i_0}}(x)\iota(I) - (H) \int_E f_{n,X_{i_0}} \right|$$

$$= \left| (D) \sum \left[f_{n,X_{i_0}}(x)\iota(I) - (H) \int_I f_{n,X_{i_0}} \right] \right|$$

$$\leq (D_1) \sum \left| f_n(x)\iota(I) - (H) \int_I f_{n,X_{i_0}} \right|$$

$$< \varepsilon.$$

We have thus proved that the sequence $\{f_n \chi_{X_{i_0}}\}$ is H-equiintegrable on E where i_0 is arbitrary and the proof is complete □

To prove the next lemma and then the final result, we need the following definition.

Definition 5.7 Let V be a measurable subset of $\overline{E}$. A sequence $\{F_n\}$ of elementary-set functions is said to satisfy the local (W)-condition on V if for every $\varepsilon > 0$ there exist a gauge δ on E and a positive integer N such that for any δ-fine partial division $D = \{(I, x)\}$ of E with $x \in V$, we have

$$\left| (D) \sum [F_n(I) - F_m(I)] \right| < \varepsilon$$

whenever $n, m \geq N$.

Note that the above condition is different from, but related to, the (W)-condition (Definition 3.7) in that the former is defined on a measurable subset Y of $\overline{E}$ whereas the latter is on a sequence $\{X_i\}$ of closed subsets of $\overline{E}$.

Lemma 5.8 *Let $\{f_n\}$ be a sequence of H-integrable functions on E and for each n, let F_n denote the primitive of f_n. Suppose that the following conditions are satisfied.*

(i) *For each x in $\overline{E}$, the sequence $\{f_n(x)\}$ converges as $n \to \infty$;*

(ii) *There exists a closed subset Y of E such that $\{F_n\}$ satisfies the local (W)-condition on $\overline{E} \setminus Y$.*

Then for each $i = 1, 2, \ldots$, and for every $\varepsilon > 0$, there exists a gauge δ on E, independent of n, such that for any δ-fine partial division $D = \{(I, x)\}$ of E with associated points $x \in \overline{E} \setminus Y$, the inequality

$$\left| (D) \sum [f_n(x)\iota(I) - F_n(I)] \right| < \varepsilon$$

holds for all n.

Proof.   Let $\{U_j\}$ be a monotone increasing sequence of open sets whose union is $\overline{E} \setminus Y$ and let $\varepsilon > 0$ be given. Since $\{F_n\}$ satisfies the local (W)-condition on $\overline{E} \setminus Y$, for each j there exist a gauge δ_j on E and a positive integer N_j such that for any δ_j-fine partial division $D = \{(I, x)\}$ of E with $x \in \overline{E} \setminus Y$, we have

$$\left| (D) \sum [F_n(I) - F_m(I)] \right| < \frac{\varepsilon}{4 \times 2^j} \qquad (5.11)$$

whenever $n, m \geq N_j$. We may assume that $\{N_j\}_{j=1}^{\infty}$ is a strictly increasing sequence. Since each f_n is H-integrable on E, by Henstock's lemma, there is a gauge δ_n^* on E such that for any δ_n^*-fine partial division $D^* = \{(J, \xi)\}$ of E, we have

$$(D^*) \sum |f_n(\xi)\iota(J) - F_n(J)| < \frac{\varepsilon}{4 \times 2^n}. \qquad (5.12)$$

We may assume that δ_m^* is finer than δ_n^* when $m > n$. Next, for each $j = 1, 2, \ldots$, let $V_j = U_j \setminus U_{j-1}$ where $U_0 := \emptyset$. Then for every $x \in V_j$ we choose $M(x) = N_{k(j)}$ where $k(j) \geq j$ such that whenever $m, n \geq M(x)$ we have

$$|f_m(x) - f_n(x)| < \frac{\varepsilon}{4 \times \iota(E)}. \qquad (5.13)$$

Now define a gauge δ on E given by

$$\delta(x) = \min(\delta^*_{M(x)}(x), \delta_{k(j)}(x))$$

when $M(x) = N_{k(j)}$. Let $D = \{(I, x)\}$ be a δ-fine partial division of E with $x \in \overline{E} \setminus Y$. For a fixed positive integer n, split the partial division D into D_1 and D_2 for which $M(x) \geq n$ and $M(x) < n$ respectively. Consequently, using inequalities (5.11), (5.12) and (5.13) we obtain

$$\left| (D) \sum [f_n(x)\iota(I) - F_n(I)] \right|$$
$$\leq \left| (D_1) \sum [f_n(x)\iota(I) - F_n(I)] \right|$$
$$+ \left| (D_2) \sum \left[f_n(x)\iota(I) - f_{M(x)}(x)\iota(I) \right] \right|$$
$$+ \left| (D_2) \sum \left[f_{M(x)}(x)\iota(I) - F_{M(x)}(I) \right] \right|$$
$$+ \left| (D_2) \sum \left[F_{M(x)}(I) - F_n(I) \right] \right|$$
$$< \frac{\varepsilon}{4} + \frac{\varepsilon}{4 \times \iota(E)} \times \iota(E) + \frac{\varepsilon}{4} + \frac{\varepsilon}{4}$$
$$= \varepsilon$$

and the result follows.

$\square$

Theorem 5.7 *Let $\{f_n\}$ be a sequence of H-integrable functions on E and for each n, let F_n denote the primitive of f_n. Suppose that the following conditions are satisfied.*

(i) *For each x in $\overline{E}$, the sequence $\{f_n(x)\}$ converges as $n \to \infty$;*

(ii) *For each $i = 1, 2, \ldots$, for every closed set $Y \subseteq X_i$ and for every interval I such that $Y \subseteq I$, the sequence $\{f_n\}$ satisfies the uniform Harnack condition on Y with respect to I;*

(iii) *There exists a positive integer i_0 such that $\{F_n\}$ satisfies the local (W)-condition on $\overline{E} \setminus X_{i_0}$;*

(iv) *For each $i = 1, 2, \ldots$, and for every interval $I \subseteq X_i$, the sequence $\{F_n(I)\}$ converges as $n \to \infty$.*

Then $\{f_n\}$ is H-equiintegrable on E.

Proof. Let $\varepsilon > 0$ be given. By Lemmas 5.7 and 5.8, and hypothesis (ii), we choose a gauge δ on E, which is independent of n and satisfies the condition that $B(x, \delta(x)) \cap X_{i_0} = \emptyset$ whenever $x \in \overline{E} \setminus X_{i_0}$, such that for any δ-fine division $D = \{(I, x)\}$ of E, we have

$$\left| (D) \sum_{x \in X_{i_0}} f_n(x)\iota(I) - (H) \int_{X_{i_0}} f_n \right| < \frac{\varepsilon}{3} \qquad (5.14)$$

and

$$\left| (D) \sum_{x \in \overline{E} \setminus X_{i_0}} [f_n(x)\iota(I) - F_n(I)] \right| < \frac{\varepsilon}{3}. \qquad (5.15)$$

Furthermore, there exists a sequence $\{A_n\}$ of real numbers such that for any nonabsolute subset E_1 of $E \setminus X_{i_0}$ involving δ, we have

$$\left| (H) \int_{E_1} f_n - A_n \right| < \frac{\varepsilon}{3}. \qquad (5.16)$$

Note that since $\{f_n\}$ satisfies the uniform Harnack condition on X_{i_0} with respect to E, by Cauchy–Harnack extension (Theorem 5.1), we have

$$A_n = (H) \int_{E \setminus X_{i_0}} f_n$$

for each n. Now let $D = \{(I, x)\}$ be a δ-fine division of E. We decompose the division D into D_1 and D_2 such that D_1 is tagged at X_{i_0} and D_2 at $\overline{E} \setminus X_{i_0}$. Define

$$E_0 = \bigcup \{I : (I, x) \in D_2\}$$

and note that E_0 is a nonabsolute subset of $E \setminus X_{i_0}$ involving δ. Then applying inequalities (5.14), (5.15) and (5.16) yields

$$\left| (D) \sum f_n(x) \iota(I) - (H) \int_E f_n \right|$$

$$\leq \left| (D_1) \sum f_n(x) \iota(I) - (H) \int_{X_{i_0}} f_n \right|$$

$$+ \left| (D_2) \sum f_n(x) \iota(I) - (H) \int_{E \setminus X_{i_0}} f_n \right|$$

$$\leq \left| (D_1) \sum f_n(x) \iota(I) - (H) \int_{X_{i_0}} f_n \right|$$

$$+ \left| (D_2) \sum f_n(x) \iota(I) - (H) \int_{E_0} f_n \right|$$

$$+ \left| (H) \int_{E_0} f_n - A_n \right|$$

$$< \varepsilon.$$

Hence $\{f_n\}$ is H-equiintegrable on E.   □

With the above result, we conclude our discussion in this book.

Bibliography

[1] S. I. Ahmed and W. F. Pfeffer, *A Riemann Integral in a Locally Compact Hausdorff Space*, J. Austral. Math. Soc., Ser. A, **41** (1986), 115–137.

[2] B. Bongiorno and L. Di Piazza, *Convergence Theorems for Generalized Riemann-Stieltjes Integrals*, Real Analysis Exchange **17**(1) (1991–92), 339–361.

[3] Z. Buczolich, *Henstock Integrable Functions are Lebesgue Integrable on a Portion*, Pro. Amer. Math. Soc. **111**(1), (Jan 1991), 127–129.

[4] P. S. Bullen, *The Burkill Approximately Continuous Integral*, J. Austral. Math. Soc., Ser. A, **35** (1983), 236–253.

[5] R. O. Davies and Z. Schuss, *A Proof that Henstock's Integral Includes Lebesgue's*, J. London Math. Soc. **2**(3) (1970), 561–562.

[6] N. Dinculeanu, *Vector Measures*, Pergamon Press, Berlin, 1967.

[7] C. S. Ding and P. Y. Lee, *Controlled Convergence Theorem for the Henstock Integral in Division Spaces* (in Chinese), Acta Mathematica Sinica **37** (1994), 497–506.

[8] A. G. Djvarsheishvili, *On a Sequence of Integrals in the Sense of Denjoy*, Akad. Nauk Gruzin. SSR Trudy Mat. Inst. Rajmadze **18** (1951), 221–236.

[9] J. Dugundji, *Topology*, Allyn and Bacon, Inc., Boston, Mass., 1967.

[10] L. Evans and R. Gariepy, *Measure Theory and Fine Properties of Functions*, CRC Press, 1992.

[11] R. A. Gordon, *Another Look at a Convergence Theorem for the Henstock Integral*, Real Analysis Exchange **15**(2) (1989–90), 724–728.

[12] R. Henstock, *Linear Analysis*, Butterworths, London, 1968.

[13] R. Henstock, *Lectures on the Theory of Integration*, World Scientific, Singapore, 1988.

[14] R. Henstock, *The General Theory of Integration*, Clarendon Press, Oxford, 1991.

[15] R. Henstock, *Measure Spaces and Division Spaces*, Real Analysis Exchange **19**(1) (1993–94), 121–128.

[16] E. Hewitt and K. A. Ross, *Abstract Harmonic Analysis*, Vol 1, 2nd Edition, Springer-Verlag, Berlin-Heidelberg-New York, 1979.

[17] E. Hewitt and K. Stromberg, *Real and Abstract Analysis*, Springer-Verlag, New York, 1969.

[18] J. Jarnik and J. Kurzweil, *Equi-integrability and Controlled Convergence of Perron-type Integrable Functions*, Real Analysis Exchange **17**(1) (1991–92), 110–139

[19] D. S. Kurtz and C. W. Swartz, *Theories of Integration, the Integrals of Riemann, Lebesgue, Henstock–Kurzweil, and McShane*, Series in Real Analysis Vol. 9, World Scientific, Singapore, 2004.

[20] J. Kurzweil, *Nichtabsolut Konvergente Integrale*, Teubner-Texte zur Mathematik, Teubner Verlagsgesellschaft, Leipzig, 1980.

[21] P. Y. Lee, *Lanzhou Lectures on Henstock Integration*, World Scientific, Singapore, 1989.

[22] P. Y. Lee, *On ACG* Functions*, Real Analysis Exchange **15**(2) (1989–90), 754–759.

[23] P. Y. Lee, *Generalized Convergence Theorems for Denjoy–Perron Integrals*, New Integrals, Lecture Notes in Math. 1419, Springer-Verlag, Berlin-Heidelberg-New York, 1990, 97–109.

[24] P. Y. Lee, *Harnack Extension for the Henstock Integral in the Euclidean Space*, J. Math. Study, **27**(1) (1994), 5–8.

[25] P. Y. Lee, *Measurability and the Henstock Integral*, Proc. International Math. Conference 1994, Kaohsiung, World Scientific, Singapore, 1995, 99–106.

[26] P. Y. Lee and T. S. Chew, *A Better Convergence Theorem for Henstock Integrals*, Bull. London Math. Soc. **17** (1985), 557–564.

[27] P. Y. Lee and T. S. Chew, *Integration of Highly Oscillatory Functions in the Plane*, Proc. Asian Math. Conference 1990, Hong Kong, World Scientific, Singapore, 1992, 277–279.

[28] P. Y. Lee and W. L. Ng, *The Radon–Nikodým Theorem for the Henstock Integral in Euclidean Space*, Real Analysis Exchange **22**(2) (1996–97), 677–687.

[29] P. Y. Lee and R. Vyborny, *Henstock–Kurzweil Integration and the Strong Lusin Condition*, Bollettino dell'Unione Matematica Italiana **7B** (1993), 761–773.

[30] P. Y. Lee and R. Vyborny, *The Integral: An Easy Approach after Kurzweil and Henstock*, Australian Mathematical Society Lecture Series 14, Cambridge Univ. Press., Cambridge, 2000.

[31] G. Q. Liu, *On Necessary Conditions for Henstock Integrability*, Real Analysis Exchange **18**(2) (1992–93), 522–531.

[32] S. P. Lu and P. Y. Lee, *Globally Small Riemann Sums and the Henstock Integral*, Real Analysis Exchange **16**(2) (1990–91), 537–545.

[33] J. Mortensen, *Advances in Geometric Integration*, Real Analysis Exchange **19**(2) (1993–94), 358–393.

[34] P. Muldowney, *A General Theory of Integration in Function Spaces*, Pitman Research Notes in Math. 153, Longman, Harlow, 1987.

[35] W. L. Ng, *A Nonabsolute Integral on Measure Spaces that Includes the Davies–McShane Integral*, New Zealand J. Math., **30** (2001), 147–155.

[36] W. L. Ng, *The Radon–Nikodým Theorem for a Nonabsolute Integral on Measure Spaces*, Bull. Korean Math. Soc., **41**(1) (2004), 153–166.

[37] W. L. Ng and P. Y. Lee, *Nonabsolute Integral on Measure Spaces*, Bull. London. Math. Soc., **32** (2000), 34–38.

[38] W. R. Parzynski and P. W. Zipse, *Introduction to Mathematical Analysis*, International Series in Pure and Applied Mathematics, McGraw-Hill, New York, 1987.

[39] W. F. Pfeffer, *The Multidimensional Fundamental Theorem of Calculus*, J. Austral. Math. Soc. **43**(2) (1987), 143–170.

[40] W. F. Pfeffer, *The Riemann Approach to Integration: Local Geometric Theory*, Cambridge University Press, Cambridge, 1993.

[41] S. Saks, *Theory of the Integral*, 2nd revised edition, Dover, New York, 1937.

[42] B. Soedijono, P. Y. Lee and T. S. Chew, *The Kubota Integral and Beyond*, NUS Research Report No. 389, May 1989.

[43] P. J. Wang, *Equi-integrability and Controlled Convergence for the Henstock Integral*, Real Analysis Exchange **19**(1) (1993–94), 236–241.

[44] P. J. Wang and P. Y. Lee, *Every Absolutely Henstock Integrable Function is McShane Integrable*, Journal of Mathematical Study (Xiamen, China) **27** (1994), 47–51.

[45] N. Wiener, *Generalised Harmonic Analysis*, Acta Math **55** (1930), 117–258.

Glossary

Absolute Continuity on E with Respect to ι A function F defined on the set of all measurable subsets of $\overline{E}$ is absolutely continuous on E with respect to ι if for every $\varepsilon > 0$, there exists $\eta > 0$ such that for every measurable subset Y of $\overline{E}$ satisfying the condition that $\iota(Y) < \eta$, we have

$$|F(Y)| < \varepsilon.$$

Absolutely Bounded Riemann Sum ($\boldsymbol{ABRS}$) Let $Y \subseteq \overline{E}$ be a measurable set. An elementary-set function F is said to have an absolutely bounded Riemann sum on Y, or simply said to be $ABRS$ on Y, if there exist a real number c and a gauge G on E such that for any G-fine division $D = \{(I, x)\}$ of E, we have

$$(D) \sum_{x \in Y} |F(I)| \le c.$$

$\boldsymbol{AC}$ An elementary-set function F is said to be AC on E if for every $\varepsilon > 0$, there exists $\eta > 0$ such that for any partial division $D = \{(I, x)\}$ of E satisfying the condition that $(D) \sum \iota(I) < \eta$, we have

$$(D) \sum |F(I)| < \varepsilon.$$

(AC)-Condition Let f be a function on $\overline{E}$ and Y be a closed subset of $\overline{E}$. We say that the function f satisfies the (AC)-condition on Y if f is H-integrable on every measurable subset of Y and for every $\varepsilon > 0$, there exists $\eta > 0$ such that whenever $W \subseteq Y$ and $\iota(W) < \eta$, we have

$$(H) \int_W |f| < \varepsilon.$$

$AC_\Delta(Y)$ Let $Y \subseteq \overline{E}$ be a measurable set. An elementary-set function F is said to be $AC_\Delta(Y)$ if F is $ABRS$ on Y and for every $\varepsilon > 0$, there exist a gauge G on E and $\eta > 0$ such that for every G-fine partial division $D = \{(I, x)\}$ of E with associated points $x \in Y$ satisfying the condition that $(D) \sum \iota(I) < \eta$, we have

$$\left| (D) \sum F(I) \right| < \varepsilon.$$

$AC^\Delta(Y)$ Let $Y \subseteq \overline{E}$ be a measurable set. An elementary-set function F is said to be $AC^\Delta(Y)$ if F is $ABRS$ on Y and for every $\varepsilon > 0$, there exist a gauge G on E and $\eta > 0$ such that for any two G-fine partial divisions $D_1 = \{(I, x)\}$ and $D_2 = \{(J, y)\}$ of E with associated points $x, y \in Y$ such that D_2 is a refinement of D_1 satisfying the condition that $(D_1 \setminus D_2) \sum \iota(I) < \eta$, we have

$$\left| (D_1 \setminus D_2) \sum F(I) \right| < \varepsilon.$$

ACG_Δ If $\overline{E}$ is the union of a sequence of closed sets X_i such that F is $AC_\Delta(X_i)$ for each i, then F is said to be ACG_Δ on E. If the union of X_i is a proper subset W of $\overline{E}$, then F is said to be ACG_Δ on W.

$\boldsymbol{ACG^{\triangle}}$ If $\overline{E}$ is the union of a sequence of closed sets X_i such that F is $AC^{\triangle}(X_i)$ for each i, then F is said to be $ACG^{\triangle}$ on E. If the union of X_i is a proper subset W of $\overline{E}$, then F is said to be $ACG^{\triangle}$ on W.

Baire Category Theorem Let Y be a closed subset of $\overline{E}$ and suppose that

$$Y = \bigcup_{i=1}^{\infty} Y_i$$

where each Y_i is closed. Then there exists an integer m such that $Y_m^o \neq \emptyset$, where Y_m^o denotes the interior of Y_m.

Basic Condition Let $F, F_1, F_2, \ldots$ be elementary-set functions. The sequence $\{F_i\}$ is said to satisfy the basic condition with F if for every $\varepsilon > 0$, there is a function $M(x)$ taking integer values such that for infinitely many $m(x) \geq M(x)$ there is a gauge G on E satisfying the condition that for any G-fine division $D = \{(I, x)\}$ of E, we have

$$\left| (D) \sum F_{m(x)}(I) - F(E) \right| < \varepsilon.$$

Basic Convergence Theorem Let $f_n, n = 1, 2, \ldots$, be H-integrable on E with primitive F_n where $f_n(x) \to f(x)$ almost everywhere in $\overline{E}$ as $n \to \infty$ and let F be an elementary-set function. Then in order that f is H-integrable on E with primitive F, it is necessary and sufficient that for every $\varepsilon > 0$ there is a function $M(x)$ taking integer values such that for infinitely many $m(x) \geq M(x)$ there is a gauge G on E satisfying the condition that for any G-fine division $D = \{(I, x)\}$ of E, we have

$$\left| (D) \sum F_{m(x)}(I) - F(E) \right| < \varepsilon.$$

Basic Sequence Let f be a real-valued function defined on $\overline{E}$. A sequence $\{X_i\}$ of measurable sets with union $\overline{E}$ such that f is H-integrable on each X_i is said to be a basic sequence of f on E. The sequence is called monotone increasing if $X_i \subseteq X_{i+1}$ for each i.

Cauchy Criterion Let E be an elementary set and let f be a real-valued function on $\overline{E}$. Then f is H-integrable on E if and only if for every $\varepsilon > 0$, there exists a gauge G on E such that for all G-fine divisions $D = \{(I, x)\}$ and $D^* = \{(J, y)\}$ of E, we have

$$\left| (D) \sum f(x)\iota(I) - (D^*) \sum f(y)\iota(J) \right| < \varepsilon.$$

Condition ($*$) For every measurable set W and every $\varepsilon > 0$, there exist an open set U and a closed set Y such that $Y \subseteq W \subseteq U$ and

$$\iota(U \setminus Y) < \varepsilon.$$

Controlled Convergence Theorem Suppose that the following conditions are satisfied.

(i) $f_n(x) \to f(x)$ almost everywhere in $\overline{E}$ as $n \to \infty$ where each f_n is H-integrable on E with primitive F_n;

(ii) The sequence $\{F_n\}$ is $UACG^\Delta$ on E;

(iii) There exists an elementary-set function F on E which is finitely additive over intervals and a gauge G on E such that for any G-fine interval-point pair (I, x), we have $F_n(I) \to F(I)$ as $n \to \infty$.

Then f is H-integrable on E and

$$\lim_{n \to \infty} (H) \int_E f_n = (H) \int_E f.$$

Darboux Property A measurable subset Y of $\overline{E}$ is said to have the Darboux property if for every real number α such that $0 \leq \alpha \leq \iota(\overline{E})$, there exists a measurable set $W \subseteq \overline{E}$ satisfying $\iota(W) = \alpha$.

δ-Fine Cover Let E_0 be an elementary subset of E, Y be a closed subset of E_0 and δ be a gauge on E. A δ-fine cover of Y is the union of the intervals $I_1, I_2, \ldots, I_n$ in a δ-fine partial division $\{(I_i, x_i) : i = 1, 2, \ldots, n\}$ of E with associated points $x_i \in Y$ such that the union contains Y.

Dominated Convergence Theorem Suppose that the following conditions are satisfied.

(i) A function f is such that $f_n(x) \to f(x)$ almost everywhere in $\overline{E}$ as $n \to \infty$ where each f_n is H-integrable on E;

(ii) For all n, $g(x) \leq f_n(x) \leq h(x)$ almost everywhere in $\overline{E}$ where g and h are H-integrable on E.

Then f is H-integrable on E and

$$\lim_{n\to\infty} (H) \int_E f_n = (H) \int_E f.$$

Egoroff's Theorem Let $f_1, f_2, \ldots$ be measurable functions on $\overline{E}$. If $f_n(x) \to f(x)$ almost everywhere in $\overline{E}$ as $n \to \infty$, then for every $\eta > 0$ there is an open set U with $\iota(U) < \eta$ such that f_n converges to f uniformly on $\overline{E} \setminus U$.

Equiintegrability Theorem Let f_n, $n = 1, 2, \ldots$, be H-integrable on E where $f_n(x) \to f(x)$ everywhere in $\overline{E}$ as $n \to \infty$. If $\{f_n\}$ is H-equiintegrable on E, then f is H-integrable on E and

$$\lim_{n\to\infty} (H) \int_E f_n = (H) \int_E f.$$

Fatou's Lemma Let f_n, $n = 1, 2, \ldots$, be non-negative and H-integrable on E with $f_n(x) \to f(x)$ almost everywhere in $\overline{E}$ as $n \to \infty$. If the sequence of the integrals of $f_1, f_2, \ldots$ is bounded, then f is H-integrable on E and

$$(H) \int_E f \leq \liminf_{n \to \infty} \left((H) \int_E f_n \right).$$

Fundamental Theorem of Calculus Let $[a, b]$ be an interval on the real line $\mathbb{R}$. If $\mathcal{F}$ is a real-valued point function which is differentiable on $[a, b]$ with derivative $\mathcal{F}'$, then $\mathcal{F}'$ is Henstock–Kurzweil integrable on $[a, b]$ and

$$(H) \int_a^b \mathcal{F}' = \mathcal{F}(b) - \mathcal{F}(a).$$

Generalised Mean Convergence Theorem Let f_n, $n = 1, 2, \ldots$, be H-integrable on E with primitive F_n and let $\{X_i\}$ be a sequence of closed subsets of $\overline{E}$ with $\overline{E} \setminus \bigcup_{i=1}^{\infty} X_i$ being of measure zero. Suppose that the following conditions are satisfied.

(i) $f_n(x) \to f(x)$ almost everywhere in $\overline{E}$ as $n \to \infty$ and $\{F_n\}$ is USL;

(ii) There is an elementary-set function F such that for each $i = 1, 2, \ldots$ and every $\varepsilon > 0$ there exist a gauge G_i on E and a positive integer N_i satisfying the condition that for any G_i-fine partial division $D = \{(I, x)\}$ of E with $x \in X_i$ we have, for all $n \geq N_i$,

$$\left| (D) \sum [F_n(I) - F(I)] \right| < \varepsilon.$$

Then f is H-integrable on E and

$$\lim_{n \to \infty} (H) \int_E f_n = (H) \int_E f.$$

Harnack Condition Let E_0 be an elementary subset of E and Y be a closed subset of E_0. A function f on E is said to satisfy the Harnack condition on Y with respect to E_0 if f is H-integrable on every measurable subset of Y and every elementary subset of $E_0 \backslash Y$, and there exists a real number A such that the following condition is satisfied. For every $\varepsilon > 0$, there exists a gauge δ on E_0 such that for any nonabsolute subset E_1 of $E_0 \backslash Y$ involving δ, we have

$$\left| (H) \int_{E_1} f - A \right| < \varepsilon.$$

Harnack Convergence Theorem Let the functions f_1, $f_2, \ldots$ be H-integrable on E and X_1, $X_2, \ldots$ be closed sets whose union is $\overline{E}$. Suppose that the following conditions are satisfied.

(i) A function f is such that $f_n(x) \to f(x)$ almost everywhere in $\overline{E}$ as $n \to \infty$;

(ii) For each $i = 1, 2, \ldots$, for every closed set $Y \subseteq X_i$ and for every interval I such that $Y \subseteq I$, the sequence $\{f_n\}$ satisfies the uniform Harnack condition on Y with respect to I;

(iii) For each $i = 1, 2, \ldots$ and for every closed set $Y \subseteq X_i$, the sequence $\{f_n\}$ satisfies the uniform (AC)-condition on Y.

Then f is H-integrable on E and

$$\lim_{n \to \infty} (H) \int_E f_n = (H) \int_E f.$$

Harnack Extension for the H-Integral Let f be a function on $\overline{E}$ and Y be a closed subset of $\overline{E}$. Suppose that the following conditions are satisfied.

(i) f is H-integrable on Y;

(ii) f is H-integrable on every elementary subset of $E \setminus Y$;

(iii) There is a real number A satisfying the condition that for every $\varepsilon > 0$ there exists a gauge δ on E such that for any nonabsolute subset E_1 of $E \setminus Y$ involving δ, we have

$$\left| (H) \int_{E_1} f - A \right| < \varepsilon.$$

Then f is H-integrable on E and

$$(H) \int_E f = (H) \int_Y f + A.$$

Henstock's Lemma If f is an H-integrable function on E then for every $\varepsilon > 0$, there exists a gauge G on E such that for any G-fine division $D = \{(I_i, x_i) : i = 1, 2, \ldots, n\}$ of E, we have

$$\sum_{i=1}^{n} \left| f(x_i) \iota(I_i) - (H) \int_{I_i} f \right| < \varepsilon.$$

(*L*)-Condition Let $\{X_i\}$ be a sequence of measurable subsets of $\overline{E}$. An elementary-set function F is said to satisfy the (*L*)-condition on $\{X_i\}$ if for every fundamental subinterval I_0 of E and for every $\varepsilon > 0$, there is a positive integer N such that for each $i \geq N$ there exists a gauge G_i on E satisfying the condition that for every G_i-fine division $D = \{(I, x)\}$ of I_0, we have

$$\left| (D) \sum_{x \notin X_i} F(I) \right| < \varepsilon.$$

(*LG*)-Condition Let $\{X_i\}$ be a sequence of measurable subsets of $\overline{E}$. A function f is said to satisfy the (*LG*)-condition on $\{X_i\}$ if for every $\varepsilon > 0$, there is a positive integer N such that for each $i \geq N$ there exists a gauge G_i

on E satisfying the condition that for every G_i-fine division $D = \{(I, x)\}$ of E, we have

$$\left| (D) \sum_{x \notin X_i} f(x) \iota(I) \right| < \varepsilon.$$

(LL)-Condition Let $\{X_k\}$ be a sequence of measurable sets whose union is $\overline{E}$. A sequence $\{F_k\}$ of elementary-set functions is said to satisfy the (LL)-condition on $\{X_k\}$ if for every $\varepsilon > 0$, there exists a positive integer N such that for each $i \geq N$, there exists a gauge G_i on E satisfying the condition that for every G_i-fine division $D = \{(I, x)\}$ of E, we have

$$\left| \sum_{k=i+1}^{\infty} \left[(D_k) \sum F_k(I) \right] \right| < \varepsilon,$$

where $D_k = \left\{ (I, x) \in D : x \in X_k \setminus \bigcup_{j=1}^{k-1} X_j \right\}$ for $k \geq i + 1$.

Local (W)-Condition Let Y be a measurable subset of $\overline{E}$. A sequence $\{F_n\}$ of elementary-set functions is said to satisfy the local (W)-condition on Y if for every $\varepsilon > 0$, there exist a gauge δ on E and a positive integer N such that for any δ-fine partial division $D = \{(I, x)\}$ of E with associated points $x \in Y$, and whenever $n, m \geq N$, we have

$$\left| (D) \sum [F_n(I) - F_m(I)] \right| < \varepsilon.$$

Mean Convergence Theorem Let W be a measurable subset of $\overline{E}$. Suppose that $f_n(x) \to f(x)$ almost everywhere in $\overline{E}$ as $n \to \infty$ where f_n, $n = 1, 2, \ldots$, is H-integrable on E and satisfies the condition that

$$(H) \int_W |f_n - f_m| \to 0 \quad \text{as } n, m \to \infty.$$

Then f is H-integrable on W and

$$(H) \int_W f_n \to (H) \int_W f \quad \text{as } n \to \infty.$$

Monotone Convergence Theorem Let f_n, $n = 1, 2, \ldots$, be H-integrable on E. If $f_n(x) \to f(x)$ as $n \to \infty$ for almost all x in $\overline{E}$ where $f_1(x) \leq f_2(x) \leq \cdots$ for almost all x in $\overline{E}$, and $\lim\limits_{n \to \infty} (H) \int_E f_n < +\infty$, then f is H-integrable on E and

$$\lim_{n \to \infty} (H) \int_E f_n = (H) \int_E f.$$

Nonabsolute Subset Let E_0 be an elementary subset of E and U be an open subset of E_0. An elementary set E_1 is called a nonabsolute subset of U if there exists a gauge δ on E such that E_1 is the complement of a δ-fine cover of $E_0 \setminus U$ relative to E_0. We say that E_1 is a nonabsolute subset of U involving δ with respect to E_0. Where there is no need to specify the gauge δ and the elementary set E_0, we simply call all such sets nonabsolute subsets of U.

Radon–Nikodým Theorem for the H-Integral Let F be an elementary-set function which is finitely additive over subintervals of E and is strongly ACG_Δ on E such that its derived sequence satisfies with F the basic condition. Then there exists a function f which is H-integrable on E such that

$$F(E_0) = (H) \int_{E_0} f$$

for all fundamental subsets E_0 of E. Moreover, f is unique in the sense that if g is any H-integrable function on E for which the above equality holds with f replaced by g, then $f = g$ almost everywhere in $\overline{E}$.

Radon–Nikodým Theorem for the HK-Integral Let F be a finitely additive interval function which is strongly ACG_Δ on E, where E is an elementary set in $\mathbb{R}^n$, such that its derived sequence satisfies with F the basic condition. Then there exists a HK-integrable function f defined on E such that

$$F(E_0) = (HK) \int_{E_0} f \, d\nu$$

for all fundamental subsets E_0 of E. Moreover, f is unique in the sense that, if g is any HK-integrable function on E for which the above equality holds with f replaced by g, then $f = g$ almost everywhere in $\overline{E}$.

Radon–Nikodým Theorem for the Lebesgue Integral Let F be a non-negative function defined on the set of all measurable subsets Y of $\overline{E}$ which is finitely additive over measurable sets and is absolutely continuous on E with respect to ι. Then there exists a non-negative function f which is Lebesgue integrable on E such that for any measurable subset Y of $\overline{E}$, we have

$$F(Y) = (L) \int_Y f.$$

Strong Lusin Condition (SL) An elementary-set function F is said to satisfy the strong Lusin condition, or briefly F is SL, if for every $\varepsilon > 0$ and every set S of measure zero, there exists a gauge G on E such that for any G-fine partial division $D = \{(I, \xi)\}$ of E with $\xi \in S$, we have

$$(D) \sum |F(I)| < \varepsilon.$$

Strongly ACG_Δ on E An elementary-set function F is said to be strongly ACG_Δ on E if there exist measurable sets $X_1, X_2, \ldots$ whose union is $\overline{E}$ such that F is $AC_\Delta(X_i)$ for each i, and if F satisifies the (L)-condition on $\{X_i\}$.

$\boldsymbol{UAC_\Delta(Y)}$ Let $\{F_n\}$ be a sequence of elementary-set functions and let $Y \subseteq \overline{E}$ be a measurable set. We say that the sequence $\{F_n\}$ is $UAC_\Delta(Y)$ if F_n is $AC_\Delta(Y)$ uniformly in n, that is, the gauge G in the definition of $AC_\Delta(Y)$ is independent of n.

$\boldsymbol{UAC^\Delta(Y)}$ Let $\{F_n\}$ be a sequence of elementary-set functions and let $Y \subseteq \overline{E}$ be a measurable set. We say that the sequence $\{F_n\}$ is $UAC^\Delta(Y)$ if F_n is $AC^\Delta(Y)$ uniformly in n, that is, the gauge G in the definition of $AC^\Delta(Y)$ is independent of n.

$\boldsymbol{UACG_\Delta}$ Let $\{F_n\}$ be a sequence of elementary-set functions. If $\overline{E}$ is the union of a sequence of closed sets X_i such that $\{F_n\}$ is $UAC_\Delta(X_i)$ for each i, then $\{F_n\}$ is said to be $UACG_\Delta$ on E. If the union of X_i is a proper subset W of $\overline{E}$, then $\{F_n\}$ is said to be $UACG_\Delta$ on W.

$\boldsymbol{UACG^\Delta}$ Let $\{F_n\}$ be a sequence of elementary-set functions. If $\overline{E}$ is the union of a sequence of closed sets X_i such that $\{F_n\}$ is $UAC^\Delta(X_i)$ for each i, then $\{F_n\}$ is said to be $UACG^\Delta$ on E. If the union of X_i is a proper subset W of $\overline{E}$, then $\{F_n\}$ is said to be $UACG^\Delta$ on W.

Uniform (AC)-Condition Let $\{f_n\}$ be a sequence of functions on E and Y be a closed subset of $\overline{E}$. We say that the sequence $\{f_n\}$ satisfies the uniform (AC)-condition on Y if each f_n is H-integrable on every measurable set $W \subseteq Y$ and for every $\varepsilon > 0$ there exists $\eta > 0$, independent of n, and a positive integer N such that whenever $W \subseteq Y$ and $\iota(W) < \eta$, and for all $n \geq N$, we have

$$(H) \int_W |f_n| < \varepsilon.$$

Uniform Harnack Condition Let E_0 be an elementary subset of E and Y be a closed subset of E_0. A sequence $\{f_n\}$ of functions is said to satisfy the uniform Harnack condition on Y with respect to E_0 if each f_n is H-integrable on every measurable subset of Y and every elementary subset of $E_0 \setminus Y$, and there exists a sequence $\{A_n\}$ of real numbers such that the following condition is satisfied. For every $\varepsilon > 0$, there exists a gauge δ on E_0, independent of n, such that for any nonabsolute subset E_1 of $E_0 \setminus Y$ involving δ, and for all n, we have

$$\left| (H) \int_{E_1} f_n - A_n \right| < \varepsilon.$$

Uniform Henstock's Lemma Suppose that the sequence $\{f_n\}$ of functions is H-equiintegrable on E. Then for every $\varepsilon > 0$, there exists a gauge G on E, independent of n, such that for any G-fine partial division $D = \{(I, x)\}$ of E and for all n, we have

$$(D) \sum \left| f_n(x)\iota(I) - (H) \int_E f_n \right| < \varepsilon.$$

Uniform (LG)-Condition A sequence $\{f_n\}$ of functions is said to satisfy the uniform (LG)-condition on $\{X_i\}$ if for every $\varepsilon > 0$, there is a positive integer N, independent of n, such that for each $i \geq N$ there is a gauge G_i on E, also independent of n, satisfying the condition that for all G_i-fine divisions $D = \{(I, x)\}$ of E, we have

$$\left| (D) \sum_{x \notin X_i} f_i(x)\iota(I) \right| < \varepsilon.$$

Uniform Strong Lusin Condition (USL) A   sequence $\{F_n\}$ of elementary-set functions is said to satisfy the uniform strong Lusin condition, or briefly, $\{F_n\}$ is USL, if for every $\varepsilon > 0$ and every set S of measure zero, there exists a gauge G on E, independent of n, such that for any G-fine partial division $D = \{(I, \xi)\}$ of E with $\xi \in S$ and for all n, we have

$$(D) \sum |F_n(I)| < \varepsilon.$$

(W)-Condition Let $\{X_i\}$ be a sequence of closed subsets of $\overline{E}$. A sequence $\{F_n\}$ of elementary-set functions is said to satisfy the (W)-condition on $\{X_i\}$ if for each $i = 1, 2, \ldots$ and every $\varepsilon > 0$, there exist a gauge G_i on E and a positive integer N_i satisfying the condition that for any G_i-fine partial division $D = \{(I, x)\}$ of E with $x \in X_i$ and whenever $n, m \geq N_i$, we have

$$\left| (D) \sum [F_n(I) - F_m(I)] \right| < \varepsilon.$$

Weakly $AC_\Delta(Y)$ Let F be an elementary-set function and let $Y \subseteq \overline{E}$ be a measurable set. The function F is said to be weakly $AC_\Delta(Y)$ if for every $\varepsilon > 0$, there exist a gauge G on E and $\eta > 0$ such that for every G-fine partial division $D = \{(I, x)\}$ of E with associated points $x \in Y$ satisfying the condition that $(D) \sum \iota(I) < \eta$, we have

$$\left| (D) \sum F(I) \right| < \varepsilon.$$

Weakly ACG_Δ Let F be an elementary-set function and let $\{X_i\}$ be a sequence of closed sets whose union is $\overline{E}$. The function F is said to be weakly ACG_Δ on E if it is weakly

$AC_\Delta(X_i)$ for each i. If the union of X_i is a proper subset W of $\overline{E}$, then F is said to be weakly ACG_Δ on W.

Weakly $UAC_\Delta(Y)$ Let $\{F_n\}$ be a sequence of elementary-set functions and let $Y \subseteq \overline{E}$ be a measurable set. The sequence $\{F_n\}$ is said to be weakly $UAC_\Delta(Y)$ if F_n is weakly $AC_\Delta(Y)$ uniformly in n, that is, the gauge G in the definition of weakly $AC_\Delta(Y)$ is independent of n.

Weakly $UACG_\Delta$ Let $\{F_n\}$ be a sequence of elementary-set functions and let $\{X_i\}$ be a sequence of closed sets whose union is $\overline{E}$. The sequence $\{F_n\}$ is said to be weakly $UACG_\Delta$ on E if it is weakly $UAC_\Delta(X_i)$ for each i. If the union of X_i is a proper subset W of $\overline{E}$, then $\{F_n\}$ is said to be weakly $UACG_\Delta$ on W.

Index

Printed in the United States
By Bookmasters